AF614738

Adenosine Triphosphate in Health and Disease

Edited by Gyula Mozsik

Published in London, United Kingdom

IntechOpen

Supporting open minds since 2005

Adenosine Triphosphate in Health and Disease
http://dx.doi.org/10.5772/intechopen.73455
Edited by Gyula Mozsik

Part of IntechOpen Book Series: Physiology, Volume 3
Book Series Editor: Angel Catala

Contributors
Ayrat U. Ziganshin, Sergey Grishin, Francisco Gabriel Vázquez-Cuevas, Mauricio Díaz Muñoz, Anaí Del Rocio Campos-Contreras, Ana Patricia Juárez-Mercado, Erandi Velázquez-Miranda, Tao Lu, Matthew Martin, Antja-Voy Hartley, Jiamin Jin, Emily Sun, Yo Nakamura, Yuriko Ban, Yoko Ikeda, Hisayo Higashihara, Shigeru Kinoshita, Gyula Mozsik

First published in London, United Kingdom, 2019 by IntechOpen
IntechOpen is the global imprint of INTECHOPEN LIMITED, registered in England and Wales, registration number: 11086078, The Shard, 25th floor, 32 London Bridge Street
London, SE19SG - United Kingdom
Printed in Croatia

British Library Cataloguing-in-Publication Data
A catalogue record for this book is available from the British Library

Additional hard copies can be obtained from orders@intechopen.com

Adenosine Triphosphate in Health and Disease
Edited by Gyula Mozsik
p. cm.
Print ISBN 978-1-83880-203-5
Online ISBN 978-1-83880-204-2
ISSN 2631-8261

For EU product safety concerns:
IN TECH d.o.o., Prolaz Marije Krucifikse Kozulić 3, 51000 Rijeka, Croatia, info@intechopen.com or visit our website at intechopen.com.

IntechOpen Book Series
Physiology
Volume 3

Gyula Mózsik, MD, PhD, ScD, is a professor emeritus of medicine in the 1st Department of Medicine, University of Pécs, Hungary. He served as the head of the department from 1993 to 2003. Dr. Mózsik specializes in medicine, gastroenterology, clinical pharmacology, and clinical nutrition. His research interests include biochemical and molecular pharmacological studies of the gastrointestinal tract, clinical pharmacology and nutrition, clinical genetic studies, and innovative pharmacological and nutritional (dietetic) research in new drug and food productions. He has published around 360 peer-reviewed papers, 196 book chapters, 692 abstracts, 19 monographs, and 32 edited books. He has also organized 38 national and international (in Croatia, France, Romania, Italy, United States, and Japan) congresses and symposia. He received an Andre Roberts award in 2014 from the International Union of Pharmacology, Gastrointestinal Section. Fourteen of his students have been appointed as full professors in Cuba, Egypt, and Hungary.

Editor of Volume 3: Gyula Mózsik
First Department of Medicine,
University of Pécs, Hungary

Book Series Editor: Angel Catala
National University of La Plata, Argentina

Scope of the Series

Modern physiology requires a comprehensive understanding of the integration of tissues and organs throughout the mammalian body, including the expression, structure, and function of molecular and cellular components. While a daunting task, learning is facilitated by our identification of common, effective signaling pathways employed by nature to sustain life. As a main example, the cellular interplay between intracellular Ca2 increases and changes in plasma membrane potential is integral to coordinating blood flow, governing the exocytosis of neurotransmitters and modulating genetic expression. Further, in this manner, understanding the systemic interplay between the cardiovascular and nervous systems has now become more important than ever as human populations age and mechanisms of cellular oxidative signaling are utilized for sustaining life. Altogether, physiological research enables our identification of clear and precise points of transition from health to development of multi-morbidity during the inevitable aging process (e.g., diabetes, hypertension, chronic kidney disease, heart failure, age-related macular degeneration; cancer). With consideration of all organ systems (e.g., brain, heart, lung, liver; gut, kidney, eye) and the interactions thereof, this Physiology Series will address aims of resolve (1) Aging physiology and progress of chronic diseases (2) Examination of key cellular pathways as they relate to calcium, oxidative stress, and electrical signaling & (3) how changes in plasma membrane produced by lipid peroxidation products affects aging physiology

Contents

Preface

This book is a collection of papers written by clinical researchers in the field, with chapters that focus on the role of adenosine triphosphate (ATP) in health and disease. It aims to fill a gap in the literature and provide more information on the regulatory mechanisms of living organisms. The book contains three excellent papers and one human clinical observation in ophthalmology.

As a clinician, I have spent the last several decades conducting research in clinical pharmacology, with a particular focus on peptic ulcer disease. As such, I introduce the book with a bit about my studies and a discussion of ATP and its relation to both healthy and ulcerated GI mucosa.

Further chapters address topics such as temperature-dependent effects of ATP on smooth and skeletal muscles, purinergic signaling as a new regulator of ovarian function, phosphorylation of NF-kB in cancer, and the role of ATP in asthenopia.

This book is designed for researchers and clinicians alike, and I hope it helps the reader to better understand the essential points in human health and disease.

Gyula Mózsik, MD, PhD, ScD (Med)
Professor Emeritus of Medicine,
First Department of Medicine,
University of Pécs, Hungary

Chapter 1

Introductory Chapter: From Adenosine Triphosphate to Basic and Clinical Research in Light of First and Second Messenger Systems to Cellular Energetical and Other Regulatory Functions of Cells in Animals and in Humans (with a Sample of Peptic Ulcer Disease Research)

Gyula Mózsik

1. Introduction

The living animals and human organisms, organs, and cells are in a good equilibrium under the normal conditions. This excellent equilibrium can be kept with a lot of regulatory mechanisms at the level of whole organisms, different organs, and different cells, which together can organize the different regulatory steps and pathways under normal conditions for the living organisms. These regulatory mechanism systems represent a wonderful world. That is the very simple explanation for that why many people do research works in these fields and they wanted to know more and more details about this wonderful world. The researchers are working in very different fields; however, all of us want to know more and more the essential and general laws of the different regulatory mechanism systems in hoping that these new observations will help us in keeping further this wonderful world in the forthcoming future.

The researcher people are biologists, bacteriologists, animal researchers, veterinary physicians, human physicians (and related specialists, like anatomists, physiologists, biochemists, pathologists, pharmacologists, basic researchers, clinicians, etc.), agricultural researchers, etc.

Basically we want to know more and more on the functions of living organisms; however, the possible approaching pathways are very different in the science; furthermore the "science" is a permanently changed process. Firstly, we try to register the reactions (answers) of the whole organism, including physical, physiological, and psychological aspects. In other words, we will see the whole organisms at the first time; however, later we want to know more on their mechanisms involved in the different "whole" reactions. Consequently, the main research tendency turned into the microworlds from the whole organisms (e.g., biochemistry, pharmacology, etc.), and now we are at the levels just at the level a small particles of cells (like different enzymes, biochemical reactions, membrane functions, nucleic acids, and very special particles).

This book contains four (five) different excellent chapters, three of them on theories of health and diseases and one more chapter dealing with human clinical problems, which together give a nice overview on the theories of human medical practical problem.

2. First and second messenger systems

The centrally and peripherally originated neural influences (mediators), different hormones, and—during the medical treatments—different drugs reach in the serum the plasma cell membranes.

The terminology of cell membrane represents a very complicated system by using this terminology. A lot of different enzymes and receptors are located in the membranes with significantly different mechanisms (functions).

The different first messengers (hormones, mediators, drugs)—if they will not be inactivated in the serum—will meet first with the cell membrane, and they will modify the regulatory mechanisms in different extents.

The so-called sodium pump has been studied widely in the physiology. This "sodium pump" system was responsible for the keeping of equilibrium between the significant concentration gradients of sodium and potassium in the serum versus intracellularly.

There was no question that this process is an energy-dependent process; however, the details were not known.

The sodium-potassium pump was discovered by Skou (in Denmark) in the 1950s, and it was proved that this sodium-potassium pump is responsible for the so-called sodium pump (1965). The sodium-potassium pump can be worked by a membrane enzyme. This enzyme is located in the membrane, splitted the mitochondrial adenosine triphosphate (ATP) and presence of Mg^{2+}, Na^{+} and K^{+}, and this process can be inhibited by application of g strophantin (ouabain). Skou (Aarhus, Denmark) was awarded the Nobel Prize of Chemistry in 1997.

Later Sutherland (who received the Nobel Prize of Physiology or Medicine in 2001) discovered the existence of adenylate cyclase enzyme. This enzyme is also located in the cell membrane, and the same electrolytes are necessary for the function as in the case of membrane ATPase.

Consequently, it became clear that the mitochondrial adenosine triphosphate is a common substrate for both membrane ATPase and adenylate cyclase.

The breakdown of mitochondrial ATP by membrane ATPase is adenosine diphosphate, while adenylate cyclase is cyclic adenosine monophosphate (cAMP). During these processes, energy will be liberated in the cells; however their extents are different from each other, namely, its value is about two times higher in the case of adenylate cyclase than in the case of membrane ATPase. The adenosine monophosphate is a common split intracellular compound after the breakdown of ADP and cAMP.

Atkinson (1968) created a formula to express the values of the actual tissue circumstances of phosphorylation/dephosphorylation by the following method: [(ATP + 0.5 ADP)/(ATP + ADP + AMP)]. This value is equal to 1, when all adenosine compounds are in the phosphorylate form, and this value is zero, when all adenosine compounds are in the dephosphorylated form. The application of this formula is very useful in different observation circumstances.

From these very short informations, our attention has been focused in our peptic ulcer research.

The second messenger systems are very complicated in our days, which are out of our present work.

The editor of this book is a physician (internist, gastroenterologist, clinical pharmacologist). However, before the editor would turn in the clinical works, he worked a cup of years in physiological and pharmacological (molecular and biochemical pharmacological) departments. When I met—as a physician—with the patients, I registered many difficulties in their everyday medical duties, namely, the patient's treatments, and before taking of good diagnoses.

3. Molecular biochemical observations in human peptic ulcer diseases

My clinical work started in a medical department at Second Department of Medicine, University of Debrecen, Hungary (1960). I met with a lot of gastroenterological patients, who originally suffered from "classic or genuine" peptic ulcer disease (PUD) (with and without gastrointestinal (GI) bleedings). We had very limited possibilities to take diagnosis (PUD) and treatments of patients with PUD in the 1960s; however, in the forthcoming 10 years, the fiberoscopes appeared, beside the X-ray examinations. The etiological role of tissue hypoxia was suggested in the development of gastroduodenal mucosal damage in association with the increased tone (activity) of the vagus nerve. In the different European countries, the patients received atropine treatment (three times/day in doses of 0.3–0.9 orally or 0.5–1.0 mg intramuscularly for 3–4 weeks). The scopolamine was used in the USA beside the atropine. Following the medical treatment, we believed that the patients healed, or we (internists) offered further the patient to surgeons for taking gastric surgery (partial gastrectomy or partly surgical vagotomy).

The increased gastric secretory acid secretion was believed to be in the background of PUD; however no objective method(s) was (were) in the hand of clinicians to measure the quantities of gastric acid secretion at that time. In the 1960s—near their end—different methods were established to measure gastric acid secretion (including the gastric basal acid outputs, BAO, and maximal acid output, MAO) using nasogastric tubes, and patients were given different doses of histamine or pentagastrin. Independently from the presence of these methods, practically these were not used generally in the everyday medical practice.

Many things (methods of clinical observations, modern endoscopic instruments) changed in the forthcoming time, and the basic pharmacological research produced a lot of tertiary and quaternary ammonium components (as antisecretory agents). The effects of these compounds were tested practically in animal observations, and these results were accepted by the clinicians and introduced into the medical treatment. The results differed in patients from those obtained in animal experiments, and in some cases no beneficial effect of used drug is obtained in patients. Between the years 1960 and 1970, a modern methodology was elaborated by us to objectively measure drug absorption, metabolism, excretion, serum levels of applied drugs, gastric acid secretory responses (BAO, MAO), parotid secretory responses, gastric motility, and gastric emptying. These results obtained in oral or parenteral application of different drugs offered a possibility to establish a complex clinical pharmacological methodology for parasympatholytics and for other drugs.

The human clinical pharmacology developed further as produced the necessary the controlled clinical pharmacological trials of different drugs. The first step was to prove that really the different drugs have any beneficial effect of the target organ. The identification and determination of drug action ("without giving any active drugs") were one of the biggest problems (placebo effect) of the human medical therapy (some to up to now, the correct understanding of this problems represents an essential scientific and medical treatment problems) (like production of new potentially new molecule is the research medicine, and the same in the clinical practice). This problem is now out of our point of view in the practical every day medical treatments.

We had good opportunity to carry out enough human studies on PUD patients with different parasympatholytics, histamine H2 receptor blockings, antigastrin, and proton pump inhibitors (PPI), antioxidants, and scavengers (so-called cytoprotective agents, like vitamin A, beta-carotene).

Just a small explanation is needed to understand the following step in our research. The vitamin A and beta-carotene are both together, and these agents are essential participants of human nutrition; however, these compounds have no any inhibitory effect on the human (and animal) gastric acid secretion. These observations excluded, at least in some part, that only the gastric acid (HCl) secretion is a responsible etiological factor to the development of GI ulcer diseases. Our clinical pharmacological studies clearly indicated existence of this phenomenon of this factor in patients with duodenal and gastric peptic ulcer. It was also very surprising when we carried out regular treatment with atropine in duodenal ulcer patients, we found that the ulcers healed; however the gastric acid secretion did not change during the chronic treatment [1]; meanwhile the terminology of "cytoprotection" was introduced by André Robert et al. in 1979, and later this name was generally accepted worldwide.

Originally, we wanted to understand the etiological background of PUD and the details of our medical activities during the patients' treatments. We tried to analyze very carefully the many results of correct human pharmacological treatments, and we were not able to understand the "essential point(s)" of etiology of PUD and our treatments. Furthermore, if the presence of tissue hypoxia is important in the development of GI mucosal damage, then we have to go further in the determination of the direction of the quantities and changes of cellular energy storage molecule, namely, "adenosine triphosphate (ATP)," in the healthy and ulcerated GI mucosa. However it is also important that the determination of ATP alone is possible only with the biochemical markers of the cell reactions. Consequently, our attention forwarded to biochemical research profile. Perhaps, it is not needed to mention that this is an extremely big challenge for the physicians working in the everyday medical practice (as internists).

The measurement of ATP in the different GI mucosal tissues represents only the equilibrium between the breakdown of ATP and its resynthesis. The oxygenization of mucosal tissues is not necessary in the case of ATP breakdown; however energy will be liberated by this biochemical reaction. The liberated energy is necessary for the normal functions of the very different cellular events (active transport at the cell membrane, secretory responses of the stomach, protein synthesis, etc.). However, the ATP resynthesis is possible only in circumstances of well-oxygenized tissues; meanwhile if the statement is true that hypoxemic event is present in the ulcerated gastrointestinal mucosa, consequently the ATP resynthesis is a priori inhibited. We suggested at the start of our biochemical examinations that we will find these results in a later time.

We had another (but essential) problem at that time, namely, in patients with PUD, we use the drugs in medical treatment; all of them block the active metabolic processes at least in the human gastric fundic mucosa (decrease of gastric acid secretion). If this statement was absolutely true, then why is the application of different drugs inhibiting the gastric acid section useful for the healing of damaged mucosal damage? There was a *contradiction in objecto*.

From 1964, we started with the biochemical examinations of the stomach in animal gastric tissues and in human gastric (fundic, antral, and jejunal) mucosa obtained in resecates at human gastric surgery. We did biochemical extractions of acid-soluble inorganic and organic phosphates, phospholipid phosphates, ribonucleic (RNA), and deoxyribonucleic acid (DNA). These biochemical fractions of gastric tissues generally represented the main components of cells: lipid (as membrane), acid-soluble inorganic and organic phosphates (mitochondrion), RNA (partly the cytoplasm and well as nucleus), and DNA (nucleus). In other words, we tried to study the different compartments of gastrointestinal cells. It was important to note that at least 0.3–0.5 g wet tissue sample was necessary to carry out of the abovementioned biochemical extractions. Of course, more tissue samples were obtained in the different parts of the human stomach, and all biochemical extractions and the classical measurements were done at the same time. The results were calculated and expressed to 1.0 mg DNA. Classical histological examinations were done. The tissue samples from the human gastric (antral), duodenal, and jejunal ulcers (after Billroth II operation) were obtained from different distances to form the edge of ulcer. In a later phase, we were able to successfully prepare membrane-bound ATP-dependent enzymes, namely, the Na^{+}-K^{+-}ATPase and adenylate cyclase, directly from the gastric mucosa from rats and animals, and we received a possibility to study actions of different drugs on both enzymes, firstly in vitro circumstances, later in living organs in animal observations (of course). These results were critically summarizing in the last years [1–4]. (I used **Tables 1, 2** and **Figures 1–5**, together with the given information texts to these demonstrations, and used in the text the reference list number used in the original summary monography).

Drugs	ATP-ADP transformation			ATP-cAMP transformation		
	Effects	Doses	(M)	Effects	Doses	(M)
Acetylcholine	Stimulation	10^{-7} M	→	Inhibition	10^{-4} M	→
Parasympatholytics:						
Atropine	Inhibition	10^{-11} M	→	Stimulation	10^{-8} M	→
Isopropamide	Inhibition	10^{-8} M	→	Stimulation	10^{-5} M	→
Gastrixon	Inhibition	10^{-8} M	→	Stimulation	10^{-4} M	→
Epinephrine	Inhibition	10^{-9} M	→	Stimulation	10^{-7} M	→
β-blocker (Visken)	Stimulation	10^{-4} M	→	Inhibition	10^{-5} M	→
Histamine	Inhibition	10^{-11} M	→	Stimulation	10^{-8} M	→
Cimetidine	Stimulation(?)	10^{-4} M	→	Inhibition	10^{-6} M	→
Pentagastrin	Inhibition	10^{-11} M	→	Stimulation	10^{-9}	→
PGE_1	Inhibition	10^{-11} M	→	Stimulation	10^{-9} M	→
PGE_2	Inhibition	10^{-11} M	→	Stimulation	10^{-9} M	→
Ouabain	Inhibition	10^{-8} M	→	Stimulation	10^{-4} M	→
cAMP	Inhibition	10^{-13} M	→			

*Direct effects of the membrane-bound ATP-splitting enzyme activities.

Table 1.
Pharmacological effects on the transformation of ATP into ADP by membrane ATPase and the ATP-cAMP transformation by adenylate cyclase from rat and human gastric, fundic, antral, duodenal, and jejunal mucosae [113] (with kind permission).*

Mediators	Actions	Affinity values (pD_2)	Intrinsic activities	
			(α)	**(pA_2)**
ATP-membrane ATP-ase-ADP				
Ach	Stimulation	5.50	1.00_{Ach}	5.50
Histamine	Inhibibion	9.70	$1.00_{Ouabain}$	9.70
Pentagastrin	Inhibibion	10.55	$0.87_{Ouabain}$	10.55
ATP-adenylate cyclase-cAMP				
Ach	Inhibition	5.30	$-0.70_{Pentagastrin}$	5.30
Histamine	Stimulation	9.30	$1.00_{Pentagastrin}$	9.30
Pentagastrin	Stimulation	9.40	1.00	9.40

Table indicates affinity (pD_2) and intrinsic activity (pA_2) curves for the actions of acetylcholine, histamine, and pentagastrin. This table also indicates the contradictory actions of these agents on Na^+-K^+-dependent and adenylyl cyclase systems.

Table 2.
Correlations between the magnitudes of drug actions on Na^+-K^+-dependent ATPase and the magnitudes of Na^+-K^+-ATPase activity prepared from the human gastric fundic mucosa.

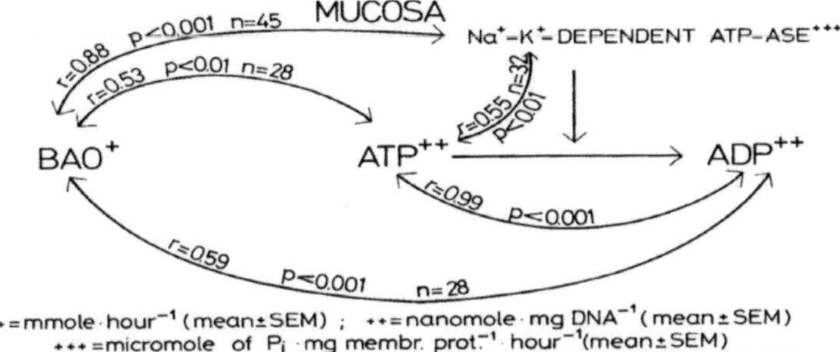

Figure 1.
Biochemically regulatory pathways between Na^+-K^+-dependent ATPase and tissue levels of ATP and ADP in the human gastric fundic mucosa on dependence of gastric BAO values (means ± SEM) [25] (with kind permission).

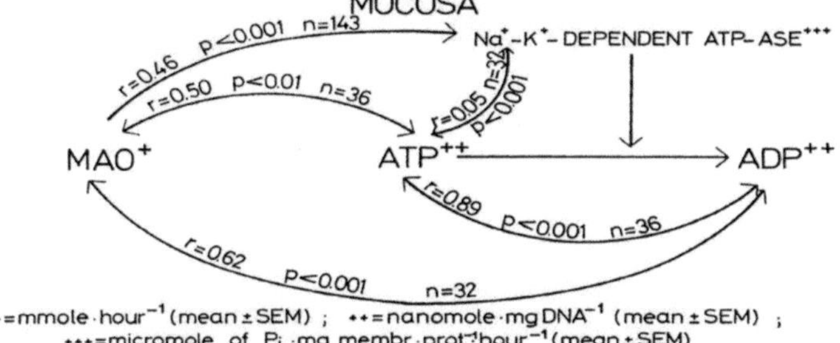

Figure 2.
Biochemically regulatory pathways between Na^+-K^+-dependent ATPase and tissue levels of ATP and ADP in the human gastric fundic mucosa in dependence of gastric maximal acid output (MAO) values [65] (with kind permission).

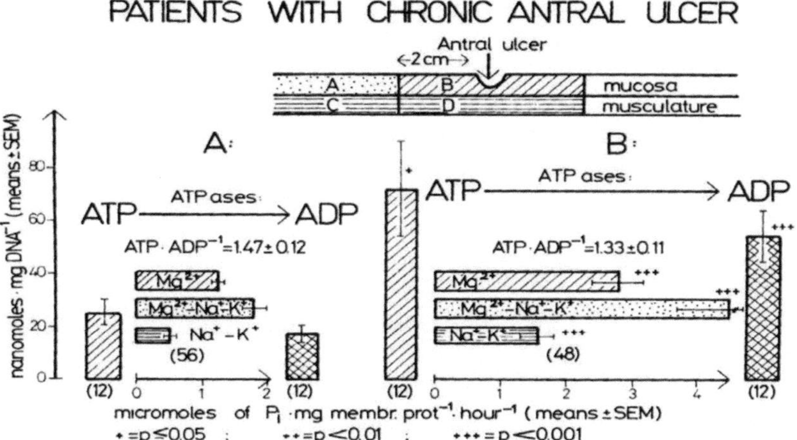

Figure 3.
Changes in the extents of ATP-ADP transformation in the antral mucosa of patients with chronic antral ulcer in the ulcerated and non-ulcerated (control) mucosae (means ± SEM) [127, 145] (with kind permission).

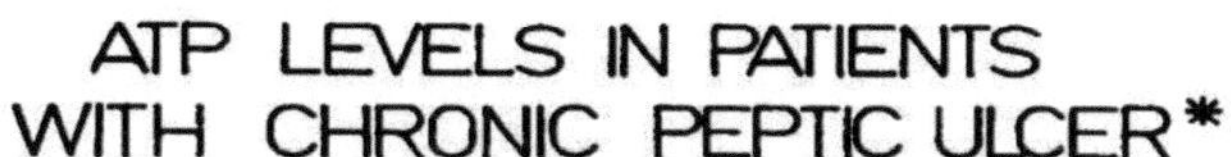

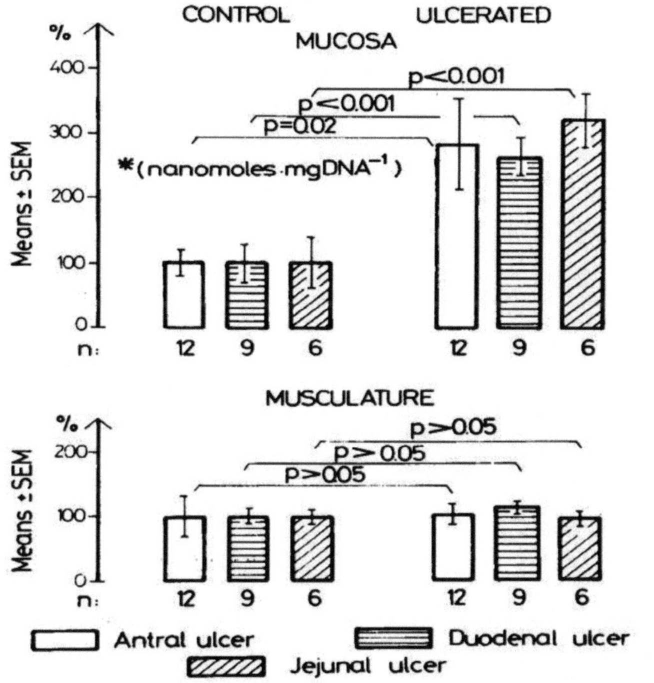

Figure 4.
Comparative demonstration in the changes of the tissue levels of ATP in the ulcerated vs. non-ulcerated antral, duodenal, and jejunal mucosae (the musculature located under the examined mucosa tissues) (means ± SEM) [143] (with kind permission).

There are some important notes from the clinical researchers and directions of basic researchers:

1. These types of biochemical pharmacological studies just are able to give an actual information (owing the human ethical positions, therapeutic protocols

used in patients' medical therapy, human rights). These problems are in the case of basic research also.

2. In animal models we really inform the provocation agents (we used 17 different models), and we used all the time a dose–response curve and followed the time-course events. All the biochemical examinations (from one animal) were done at the same time. The results were calculated. The applied agents were expressed in molecular weights, and the affinity and intrinsic values (curves) were calculated from the obtained dose–response curves.

3. The evaluation of the obtained results was evaluated at that time when the whole observations were finished.

4. These types of examination can be very hard work for clinicians.

5. I tried to demonstrate by my works how the clinicians are able to create a special bridge between the basic and clinical research works.

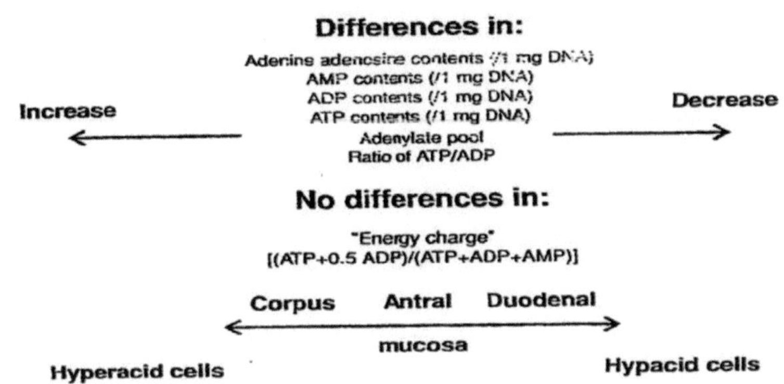

Figure 5.
The schematic presentation of biochemical buildup of human gastrointestinal mucosa in patients with different gastric acid secretory responses. All of the adenine and adenosine compounds increased significantly (meanwhile the stream of ATP breakdown enhanced in both directions) in the gastric corpus mucosa in comparison to those results obtained in the corpus mucosa of patients with hyperacidity. The same biochemical parameters were obtained in the ulcerated antral, duodenal, and jejunal mucosa in patients with chronic antral, duodenal, and jejunal mucosae. So the biochemical structure of human chronic antral, duodenal, and jejunal mucosa is the same as that obtained in the corpus fundic mucosa in patients with gastric hyperacidity [149, 150] (with kind permission).

Acknowledgements

The editor thanks the excellent cooperation of the contributors in this book during the time of this edition.

The editor is especially thankful for the given by Ms. Danijela Duric (Publishing Process Manager from Intech Open Access Publisher) and by the publisher.

Author details

Gyula Mózsik
First Department of Medicine, University of Pécs, Hungary

*Address all correspondence to: gyula.mozsik@gmail.com

References

[1] Mózsik G, Szabo IL. Membrane-bound ATP-dependent Energy Systems and the Gastrointestinal Mucosal Damage and Prevention. Rijeka: InTech Open Acces/Open Minds Publishers; 2016. pp. 1-356. DOI: 105772/60095. ISBN: 978-953-51-2251-7

[2] Mózsik G, Szabo IL, Czimmer J. Membrane-bound ATP-dependent energy system, as extra- and intracellular key signals for gastrointestinal functions and their regulations in the gastrointestinal mucosa. Current Pharmaceutical Design. 2017;**23**:1-31

[3] Mózsik G. History of the International Conferences on Ulcer Research (ICUR). Hungary: University Press of Pécs; 2006. pp. 1-228

[4] Mózsik G, Szabó S, Takeuchi J. History of the Gastrointestinal Section of the International Union of Pharmacology (IUPHAR GI Section). Hungary: University Press of Pécs; 2006. pp. 1-164

Chapter 2

Temperature-Dependent Effects of ATP on Smooth and Skeletal Muscles

Ayrat U. Ziganshin and Sergey N. Grishin

Abstract

ATP acting via different subtypes of P2X and P2Y receptors induces contractions or relaxation of mammalian smooth muscles, while in skeletal muscles, ATP can pre- and postsynaptically modulate effect of acetylcholine. It was shown that effects of ATP on both types of the muscle are significantly changed when the temperature shifts from physiological condition. For example, contractile responses of rodent urinary bladder and vas deferens mediated by P2X receptors are markedly increased with the decrease of the temperature. Similarly, in frog skeletal muscles, ATP-induced inhibition of acetylcholine release became more pronounced at low temperatures. In case of mammalian skeletal muscle, effect of temperature on ATP-induced responses depends on the type of muscle—slow and fast. In this chapter, we will discuss temperature-dependent effects of ATP on different muscle contractility and their possible mechanisms.

Keywords: ATP, P2 receptors, temperature, hypothermia, smooth muscles, skeletal muscles, contractility, neuromuscular synapse

1. Introduction

It is widely accepted now that ATP, except well-known role as an intracellular source of energy, can regulate many important cell functions acting via specific extracellular receptors, namely P2 receptors [1]. P2 receptors are divided into two families, P2X and P2Y receptors, P2X receptors being a ligand-gated ion channel, while P2Y receptors are G protein-coupled. Seven subtypes of P2X and eight subtypes of P2Y receptors are well identified and put into current classification of receptors [2, 3].

P2X and P2Y receptors are widely distributed in animal and human tissues including smooth and skeletal muscles. In smooth muscles, stimulation of P2X receptors causes contractile responses, while stimulation of P2Y receptors usually leads to relaxant effects [4]. In contrast, in adult skeletal muscles, it has been established that, while stimulation of P2 receptors does not cause either contraction or relaxation, it significantly inhibits transmitter release at the neuromuscular junction [5, 6].

Although most experiments on P2 receptors were carried out on normal temperature conditions, we have shown in our publications that in several animal smooth and skeletal muscles, the responses mediated by both P2X and P2Y receptors are

IntechOpen

significantly affected by changing the temperature conditions. In this chapter, we will review our earlier and recent studies as well as those findings done in other laboratories.

2. Guinea pig smooth muscle tissues

Our first publication on temperature dependency of the P2 receptor-mediated processes was about guinea pig smooth muscle tissue [7]. We registered responses of isolated guinea pig urinary bladder and vas deferens (P2X receptors) and taenia caeci (P2Y receptors) at the three temperature conditions of 30, 37, and 42°C. We found that the contractile responses of both urinary bladder and vas deferens to a P2X receptor agonist α,β-methylene ATP (α,β-meATP) and to electrical field stimulation in the presence of atropine and phentolamine were markedly more prominent at a temperature of 30°C than at 37 or 42°C. Similarly, relaxation of carbachol-precontracted taenia caeci caused by electrical field stimulation temperature dependently increased with decrease of temperature, while relaxation of this tissue by exogenous ATP was not affected by the temperature. A P2 receptor antagonist pyridoxalphosphate-6-azophenyl-2′,4′-disulphonic acid (PPADS) at all three temperature conditions concentration-dependently antagonized contractile responses to α,β-meATP and electrical field stimulation in both urinary bladder and vas deferens. PPADS, even at the highest concentration tested, had no effect on the relaxant responses of the taenia caeci either to electrical field stimulation or ATP, and its action was not affected by the change of temperature. It was concluded that the effectiveness of P2 receptor-mediated responses in guinea pig urinary bladder, vas deferens, and taenia caeci increases by the decrease of temperature.

Temperature dependency for some receptor-mediated responses has been tested earlier on several animal and human tissues. Using guinea pig ileum and trachea and rat vas deferens and atria preparations, hypothermia-induced supersensitivity to adenosine has been established for responses mediated via adenosine A1, but not adenosine A2, receptors [8]. It has been shown that in the rabbit central ear artery, but not femoral artery, cooling to 24°C reduces contraction, increases the relaxation caused by histamine [9], and enhances the relaxation caused by cholinoceptor stimulation [10]. Later in the study from the same laboratory, it was shown that in rabbit central ear artery at 30°C α1-adrenoceptor-mediated response is reduced and the P2 receptor-mediated component becomes more prominent [11]. On the other hand, it was found that the release of ATP from rabbit pulmonary artery induced by methoxamine, an α1-adrenoceptor agonist, being observed at 37°C, was completely eliminated at a temperature of 27°C [12].

The increase of bladder contractility at low temperature might be due to activation of cold receptors in the bladder, the presence of which has been shown both in animal and human urinary bladder [13, 14]. However, it is unlikely that cold receptors are involved in the effects which we registered in the present study since the threshold temperature to stimulate these receptors was found to be less than 30°C, and the maximum effect was registered at around 20°C [13].

It is generally accepted that in the presence of adreno- and cholinoceptor blockers, the contractions of guinea pig vas deferens and urinary bladder are mediated by P2X receptors, while in guinea pig taenia caeci, low-frequency electrical field stimulation evokes the relaxation via P2Y receptors [15–17]. We have found that both P2X and P2Y receptor-mediated responses elicited by electrical field stimulation are increased at low temperature. It could be suggested that this effect occurs due to the decrease of activity of the transmitter-metabolizing enzymes, namely, ecto-ATPase and ectonucleotidases during cooling, since it is generally accepted that ecto-ATPase activity is temperature-dependent, with the optimum temperature of 37°C for warm-blood temperature animals [18]. However, this cannot explain results with the enzymatically stable P2X receptor agonist α,β-meATP, the effects of which are not affected by ecto-ATPases. Moreover, in taenia caeci when we used ATP, which is readily degraded by ecto-ATPases, we did not find any temperature dependency in agonist activity. Thus, it seems that supersensitivity of P2 receptors at a low temperature is a feature of receptor itself and is not dependent on ecto-ATPase activity.

It was believed initially that PPADS was a selective P2X receptor antagonist [19, 20] although later antagonism of recombinant P2Y receptors by PPADS was reported [21]. In our earlier study, we established that in the guinea pig taenia caeci, substantial antagonism against P2Y receptor-mediated relaxation was obtained only at a concentration of 100 μM of PPADS [22]. Similarly, in these experiments we did not find any antagonism at P2Y receptors of PPADS at concentrations up to 30 μM on taenia caeci. Thus, it supports the view that at least in the pharmacological organ bath experiments, PPADS shows relatively good selectivity to P2X receptors.

It was an interesting finding that in taenia caeci responses to electrical field stimulation were clearly temperature-dependent, while the relaxation caused by exogenous ATP was statistically identical at different temperature conditions. Since it has been clearly shown that ATP is a transmitter which is released during electrical field stimulation of guinea pig taenia caeci to act on P2Y receptors, it seems that in this tissue only prejunctional mechanisms of transduction are sensitive to the shifts of the temperature while postjunctional processes are not.

3. Frog skeletal muscle

Next, we decided to test the P2 receptor-mediated effects in tissues of cold-blooded animals. For that the contractile responses of isolated *Rana ridibunda* frog sartorius muscle contractions evoked by electrical field stimulation (EFS) were studied at three temperature conditions of 17, 22, and 27°C [23]. ATP concentration dependently inhibited the electrical field stimulation-evoked contractions of sartorius muscle at all three temperatures; this effect was significantly more prominent at a temperature of 17°C than at the other two temperatures. Adenosine also caused inhibition of electrical field stimulation-evoked contractions which was statistically identical at all three temperature conditions tested. A P2 receptor antagonist, PPADS, reduced the inhibitory effect of ATP at all three temperatures but did not affect inhibitory action of adenosine. In contrast, 8-(p-sulfophenyl)theophylline (8-SPT), a nonselective P1 receptor antagonist, abolished inhibitory effects of adenosine at all three temperature conditions but did not antagonize inhibition caused by ATP. In electrophysiological experiments, ATP and adenosine temperature dependently reduced end-plate currents recorded in sartorius neuromuscular junction by voltage clamp technique. The inhibitory effects of both agonists were enhanced with the decrease of temperature. 8-SPT abolished the inhibitory effect of adenosine but not ATP on end-plate currents. Suramin, a nonselective P2 receptor antagonist, inhibited the action of ATP but not adenosine, while PPADS had no influence on the effects of either ATP or adenosine. It was concluded from this study that the effectiveness of P2 receptor-mediated inhibition of frog skeletal muscle contraction in contrast to that of adenosine is dependent on the temperature conditions.

Thus, we had demonstrated that presynaptic P2 receptor-mediated inhibition of the frog skeletal muscle contractions produced by nerve stimulation has a clear temperature-dependent feature—lowering the temperature leads to the increase of

P2 receptor-mediated inhibition. The depressant effect of exogenous ATP on neuromuscular transmission was demonstrated for the first time more than 40 years ago [24], although for a long time, it was believed that inhibitory action of ATP is indirect and depends on degradation to adenosine [25, 26]. In mammalian tissues, the existence of presynaptic P2 receptors at neuromuscular junction was suggested by immunohistochemical analysis [27], and electrophysiologically, it was established that ATP but not adenosine inhibited nonquantal release of acetylcholine [5]. At frog neuromuscular junction, it was shown that ATP inhibited transmitter release via presynaptic P2 receptors [6], and it was proposed that ATP produces its effect via P2Y2-like receptors coupled to multiple intracellular cascades [28].

Temperature dependency of skeletal muscle contractility is a known phenomenon. It has been shown that this phenomenon has an endothermic nature, and raising the temperature increases the force and the strain of the myosin heads attached in the isometric contraction [29]. The decrease of contractile force at lower temperature could be due to the attenuation of metabolic enzyme activities [30, 31] or processes of energy production and transfer [32, 33].

In contrast to adenosine, we have found that the effect of ATP on neuromuscular transmission was temperature-dependent in functional experiments. Lowering the temperature caused the increase of ATP-induced inhibition of electrical field stimulation-evoked contractions, and this effect was highly sensitive to P2 receptor antagonist PPADS and not sensitive to 8-SPT, a P1 antagonist. These differences between two purines are thought to be coupled at their action mechanism. Both ATP and adenosine reduce quantal release of acetylcholine [6], thereby decreasing amplitude of postsynaptic end-plate currents. However, the temperature-mediated effect of ATP is more prominent and can achieve corresponding to amplitude of end-plate current reduction of the muscle contraction.

To find the nature of receptors involved, we used PPADS, a P2 receptor antagonist with a preferential effect on P2X receptors in functional whole tissue experiments [19, 20], and found that ATP-evoked inhibition of muscle contraction was highly sensitive to this antagonist, while in electrophysiological study, it failed to affect responses to ATP. However, another nonselective P2 receptor antagonist suramin [15] significantly reduced ATP-induced inhibition. Although both PPADS and suramin are considered as nonselective P2 receptor antagonists, it has been shown that suramin, compared to PPADS, has a more broad P2 receptor antagonist activity, affecting most of P2X and P2Y receptor subtypes [34]. For instance, it has been shown that recombinant P2Y2 receptors are sensitive to antagonistic effect of suramin but not of PPADS [35]. In addition, in organ bath pharmacological experiments, PPADS tends to antagonize mostly P2X receptor subtypes [16, 19, 20], blocking P2Y receptor-mediated processes only at higher concentrations [22]. Neither PPADS nor suramin affects inhibition caused by adenosine. These results support the view that ATP inhibited the electrical field stimulation-evoked contractions of frog skeletal muscle by acting on presynaptic P2 receptors. It is most likely that these receptors belong to P2Y family, but involving some subtypes of P2X receptors cannot be ruled out at present.

It has been proposed that purine nucleotides and nucleosides were among the first neurotransmitters in the evolution and development of the living cells [36, 37]. Thus, it is possible that, in phylogenetically older animals, in which organism is functioning in low-temperature conditions, the transmitter role of purine nucleosides and nucleotides in cell-to-cell communications is as important as well-known intracellular metabolic actions of purines (production of energy, involvement in synthesis of nucleic acids). Thus, we suggest that supersensitivity of P2 receptor-mediated responses at lower temperature, which we have demonstrated in mammal

[7] and amphibian tissues [23], is a fundamental feature of these receptors which could be a reflection of their past role in the early stage of evolution.

4. Rodent skeletal muscles

When we found that, similar to that in guinea pig smooth muscle tissues, P2 receptor-mediating processes in amphibian skeletal muscles are markedly more pronounced in low-temperature condition, we did the next study using rodent skeletal muscles, namely rat soleus muscle [38].

We registered carbachol- and electric field stimulation-induced contractions of rat soleus muscle in norm and in the presence of ATP under different temperature conditions—37 and 14°C. We found that with decreasing temperature, both the force and the time of contractions are increased. ATP inhibited the amplitude of contraction caused by indirect stimulation by an electric field; in this case, the combined pre- and postsynaptic modulation effect of this purine was observed. To separate these effects, we investigated the effect of ATP on carbachol-induced contraction. In this mode, ATP increased the contraction of the "slow" muscle. With a decrease in temperature, both pre- and postsynaptic effects of ATP are enhanced, but not equivalent. The increasing potentiating effect of ATP with the use of postsynaptic P2 receptors overlaps and masks an increased, but to a lesser extent, inhibitory presynaptic effect [38].

Maintaining the body temperature in certain range provides for warm-blooded animals the ability to move and perform motor activity in a wide range of temperature differences of the environment [39].

It is known that in the absence of significant fluctuations in the temperature of the internal organs of mammals, the peripheral parts of their body can experience significant changes in temperature, for example, up to 15°C decrease in humans [40, 41]. Thus, the peripheral skeletal muscles of warm-blooded animals retain the ability to contractile activity even with a significant decrease in their temperature.

According to modern ideas, the strength, speed of contraction, and relaxation of skeletal muscles of warm-blooded animals as a rule increase with increasing temperature [42, 43]. This was observed, for example, during physical exercises when the temperature of skeletal muscles on the periphery of the human body increased by several degrees Celsius [44–46].

At the same time, the above studies did not attach special importance to the types of skeletal muscle examined. It is known that several types of phase skeletal muscle fibers are distinguished from which "slow" and "fast" are distinguished in all classifications [47, 48]. It is understandable that these muscles differing in their very function—maintaining pose ("slow" muscles) and performing subtle movements ("fast")—are made to react differently to temperature changes which is observed in practice [49–51].

Molecular non-quantum secretion of the mediator still has not given a fundamental importance due to the lack of a generalized action. Indeed, despite its large value, non-quantum secretion only depolarizes the end-plate region by ~5 mV which can be determined by hyperpolarization in the presence of postsynaptic receptor blockers—the "H effect" [52–54]. However, non-quantum secretion is extremely important and is crucial for the functioning of the synapse. It should be noted that the evaluation of non-quantum secretion of the myoneural synapse in the cold-blooded is difficult because of the small value of the registered H effect [54, 55].

The temperature dependence of the magnitude of non-quantum secretion in neuromuscular preparations of rodents is complex. It was found that in the range from 10 to 35°C, the size of non-quantum secretion has two relative peaks at 20 and 35°C and two minima at 25 and 10°C (in the latter case, the H effect is not expressed at all) [54–56].

The frequency of spontaneous single-quantum responses in synapses of warm-blooded animals increases as the temperature increases [56–58] without changing the amplitude of these responses [59]. Thus, two processes of acetylcholine release, namely quantum and non-quantum, have a different temperature dependence, which indicates the presence of independent mechanisms [55, 58].

The analysis of the published data leads to an unequivocal conclusion about the rise of the synaptic delay of postsynaptic responses with decreasing temperature on rodent preparations [60, 61]—the same as in cold-blooded.

It is known that in the neuromuscular synaptic cleft, there is an acetylcholinesterase which rapidly cleaves the neurotransmitter acetylcholine [62]. It was shown that when the temperature of the rat diaphragm preparation was reduced from 37 to 17°C, the activity of acetylcholinesterase decreased by 34% [63]. A similar pattern was observed in experiments with the preparation of the frog sartorius muscle which led to a suggestion that a decrease in the activity of acetylcholinesterase is responsible for an increase of the time course of the end-plate current at hypothermia [64].

To study the state of postsynaptic cholinergic receptors, cholinomimetics (primarily the slowly decaying cholinomimetic agent—carbachol) and cholinolytics are used. In experiments with slow muscle preparations such as rat m. soleus, the amplitude of the miniature potentials of the terminal plate did not change after application of carbachol at a temperature range from 18 to 38°C. On the other hand, at temperatures 37–38°C, there was a 40% decrease in the incidence of spontaneous postsynaptic responses in the presence of this cholinomimetic (in which combination indicates presynaptic nature of the effect) [65].

There are several studies on the temperature dependence of the contractile apparatus of "slow" skeletal muscles with conflicting data on the muscles of the same animals [66–70]. Thus, according to some sources, the temperature dependence of the "slow" muscle fibers of the rat is much more pronounced than the "fast" ones [42, 71–79], while others found that the temperature sensitivity of the myosin of the "slow" muscle fibers of the rat does not differ from the "fast ones" [80, 81].

In experiments on demembranized muscle fibers, where the temperature effects of electromechanical coupling are not relevant and only the modulation of the mechanical function itself plays a role, it was clarified that the dependence of the reduction force of "slow" and "fast" fibers on temperature is similar [80]. With an increase of temperature from 10 to 35°C, the force of contraction of "slow" fibers increased threefold and the "fast" ones by three and a half. The situation is different with the rate of contraction. Thus, the rate of reduction of "fast" fibers increased with increasing temperature from 10 to 35°C, while for "slow" fibers this parameter changed insignificantly [80].

We found that as the temperature is lowered, the force of contractions of the slow skeletal muscle of the rat increases [38], while the fast one decreases. With carbachol-induced contraction (when only receptors of the postsynaptic membrane are stimulated), as the temperature decreases, the amplitude of contractions of the slow muscle of the rat also increases [38].

It is known that a decrease in Ca^{2+} concentration, which provides exocytosis of the neurotransmitter quanta [82–84], reduces the strength of the contraction of the skeletal muscle over time—in contrast to the rapid effect on cardiomyocytes [85]. This action is temperature-dependent; in rat fast muscle fibers, the contraction force increased with increasing concentration of calcium ions; the lower the temperature, the more is the effect. A rat slow muscle did not produce similar effect [86].

5. Conclusion

Despite a large number of studies, the direct mechanisms of temperature effects on the functioning of the muscle remain unsolved [87, 88]. Moreover, if the muscle is stimulated directly, then the temperature change no longer has such an effect. That is, the reason for the phenomenon being discussed is synaptic.

It was believed that changes in motor units at low temperature are due to depletion of the activity of metabolic enzyme systems [30, 31] or energy synthesis and transfer processes [89, 90]. However, it was clear that the contribution of the temperature sensitivity of the muscle biochemical processes cannot justify such dramatic changes in the nature of the contraction of the whole muscular organ with a change in temperature [91]. We suggest that the temperature-sensitive tonic effects of endogenous ATP during the contraction can underlie the phenomenology of changes in muscle responses with decreasing temperature.

In conclusion, we believe that studying the effects of hypothermic conditions has not only theoretical significance but also potentially important clinical implications since hypothermia is widely used in clinical practice for cerebral protection during surgical interventions or resuscitation of critically ill patients [92–96]. This underlines the importance of studying the reaction of other organs and tissues to hypothermia and especially the effect that low temperatures have on receptor-based interactions. Such studies add important information regarding the activity of P2 receptors under hypothermic conditions in mammalian muscles. Although these results cannot be directly transferred to human muscle tissues, they provide important insight into how activation of human P2 receptors might behave under hypothermia and predict how effects of certain drugs might be altered by this nonphysiologic state.

Acknowledgements

This study was partly supported by the Russian Foundation for Basic Research grant nos. 16-04-00101 and 18-44-160009.

Conflict of interest

The authors declare the absence of conflict of interest.

Author details

Ayrat U. Ziganshin* and Sergey N. Grishin
Kazan State Medical University, Kazan, Russia

*Address all correspondence to: auziganshin@gmail.com

References

[1] Burnstock G. Purinergic signalling: Its unpopular beginning, its acceptance and its exciting future. BioEssays. 2012;**34**:218-225. DOI: 10.1002/bies.201100130

[2] Alexander SP, Davenport FP, Kelly E, et al. The Concise Guide to Pharmacology 2015/16: G protein-coupled receptors. British Journal of Pharmacology. 2015;**172**:5744-5869. DOI: 10.1111/bph.13348

[3] Alexander SP, Davenport FP, Kelly E, et al. The Concise Guide to Pharmacology 2015/16: Ligand-gated ion channels. British Journal of Pharmacology. 2015;**172**:5870-5903. DOI: 10.1111/bph.13350

[4] Hoyle CHV. Transmission: Purines. In: Burnstock G, Hoyle CHV, editors. Autonomic Neuroeffector Mechanisms. Chur. Harwood Academic Publishers; 1992. pp. 367-407

[5] Galkin AV, Giniatulli RA, Mukhtarov MR, et al. ATP but not adenosine inhibits nonquantal acetylcholine release at the mouse neuromuscular junction. The European Journal of Neuroscience. 2001;**13**:2047-2053

[6] Giniatullin RA, Sokolova EM. ATP and adenosine inhibit transmitter release at the frog neuromuscular junction through distinct presynaptic receptors. British Journal of Pharmacology. 1998;**124**:839-844

[7] Ziganshin AU, Rychkov AV, Ziganshina LE, Burnstock G. Temperature dependency of P2 receptor-mediated responses. European Journal of Pharmacology. 2002;**456**:107-114

[8] Broadley KJ, Broome S, Paton DM. Hypothermia-induced supersensitivity to adenosine for responses mediated via A1-receptors but not A2-receptors. British Journal of Pharmacology. 1985;**84**:407-415

[9] Fernandez N, Garcia-Villalon AL, Borbujo J, et al. Cooling effects on histaminergic response of rabbit ear and femoral arteries: Role of the endothelium. Acta Physiologica Scandinavica. 1994;**151**:441-451

[10] Monge L, Garcia-Villalon AL, Montoya JJ, et al. Role of the endothelium in the response to cholinoceptor stimulation of rabbit ear and femoral arteries during cooling. British Journal of Pharmacology. 1993;**109**:61-67

[11] Garcia-Villalon AL, Padilla J, Monge L, et al. Role of the purinergic and noradrenergic components in the potentiation by endothelin-1 of the sympathetic contraction of the rabbit central ear artery during cooling. British Journal of Pharmacology. 1997;**122**:172-178

[12] Takeuchi K, Shinozuka K, Akimoto H, et al. Methoxamine-induced release of endogenous ATP from rabbit pulmonary artery. European Journal of Pharmacology. 1994;**254**:287-290

[13] Geirson G, Lindström S, Fall M. The bladder cooling reflex in man—characteristics and sensitivity to temperature. British Journal of Urology. 1993;**71**:675-680

[14] Lindström S, Mazieres L. Effect of menthol on the bladder cooling reflex in cat. Acta Physiologica Scandinavica. 1991;**141**:1-10

[15] Hoyle CHV, Knight GE, Burnstock G. Suramin antagonizes responses to P2-purinoceptor agonists and purinergic nerve stimulation in the guinea-pig urinary bladder and taenia coli. British Journal of Pharmacology. 1990;**99**:617-621

[16] McLaren GJ, Lambrecht G, Mutschler E, et al. Investigation of the actions of PPADS, a novel P2Xpurinoceptor antagonist, in the guinea-pig vas deferens. British Journal of Pharmacology. 1994;**111**:913-917

[17] Stjarne L, Stjarne E. Electrophysiological evidence that the twitch contraction of guinea-pig vas deferens is triggered by phasic activation of P2X-purinoceptors. Acta Physiologica Scandinavica. 1997;**160**:295-296

[18] Ziganshin AU, Ziganshina LE. P2 Receptors: Perspective Targets for Future Drugs. Moscow: Geotar-Media; 2009. 136 p

[19] Ziganshin AU, Hoyle CHV, Bo X, et al. PPADS selectively antagonizes P2X-purinoceptor-mediated responses in the rabbit urinary bladder. British Journal of Pharmacology. 1993;**110**:1491-1495

[20] Ziganshin AU, Hoyle CHV, Lambrecht G, et al. Selective antagonism by PPADS at P2Xpurinoceptors in rabbit isolated blood vessels. British Journal of Pharmacology. 1994;**111**:923-929

[21] Boyer JL, Zohn IE, Jacobson KA, Harden TK. Differential effects of P2-purinoceptor antagonists on phospholipase C-coupled and adenylyl cyclase-coupled P2Y-purinoceptors. British Journal of Pharmacology. 1994;**113**:614-620

[22] Windscheif U, Pfaff O, Ziganshin AU, et al. The inhibitory action of PPADS on the relaxant responses to adenine nucleotides or electrical field stimulation in guinea-pig taenia caeci and rat duodenum. British Journal of Pharmacology. 1995;**115**: 1509-1517

[23] Ziganshin AU, Kamaliev RR, Grishin SN, et al. The influence of hypothermia on P2 receptor-mediated responses of frog skeletal muscle. European Journal of Pharmacology. 2005;**509**: 187-193

[24] Ribeiro J, Walker J. The effects of adenosine triphosphate and adenosine diphosphate on transmission at the rat and frog neuromuscular junctions. British Journal of Pharmacology. 1975;**54**:213-218

[25] Redman RS, Selinsky EM. ATP released together with acetylcholine as the mediator of neuromuscular depression at frog motor nerve endings. The Journal of Physiology. 1994;**477**:117-127

[26] Ribeiro JA, Sebastio AM. On the role, inactivation and origin of endogenous adenosine at the frog neuromuscular junctions. The Journal of Physiology. 1987;**384**: 571-585

[27] Parson SH, Knutsen PM, Deuchars J. Immunohistochemical evidence that P2X receptor is present in mammalian motor nerve terminals. The Journal of Physiology. 2000;**526**:60

[28] Sokolova E, Grishin S, Shakirzyanova A, et al. Distinct receptors and different transduction mechanisms for ATP and adenosine at the frog motor nerve endings. The European Journal of Neuroscience. 2003;**18**:1254-1264

[29] Piazzesi G, Reconditi M, Koubassova N, et al. Temperature dependence of the force-generating process in single fibers from frog skeletal muscle. The Journal of Physiology. 2003;**549**:93-106

[30] MacDonald JA, Storey KB. Protein phosphatase type-1 from skeletal muscle of the freeze-tolerant wood frog. Comparative Biochemistry & Physiology Part B: Biochemical and Molecular Biology. 2002;**131**: 27-36

[31] Cowan KJ, Storey KB. Freeze-thaw effects on metabolic enzymes in wood frog organs. Cryobiology. 2001;**43**:32-45

[32] Boutilier RG, St-Pierre J. Adaptive plasticity of skeletal muscle energetics of hibernating frogs: Mitochondrial proton leak during metabolic depression. The Journal of Experimental Biology. 2002;**205**:2287-2296

[33] Mantovani M, Heglund NC, Cavagna GA. Energy transfer during stress relaxation of contracting frog muscle fibres. The Journal of Physiology. 2001;**537**:923-939

[34] Burnstock G. Purine-mediated signalling in pain and visceral perception. Trends in Pharmacological Sciences. 2001;**22**:182-188

[35] Charlton SJ, Brown CA, Weisman GA, et al. PPADS and suramin as antagonists at cloned H2Y- and P2U-purinoceptors. British Journal of Pharmacology. 1996;**118**:704-710

[36] Burnstock G. Purinoceptors: Ontogeny and phylogeny. Drug Development Research. 1996;**39**:204-242

[37] Trams GE. On the evolution of neurochemical transmission. In: Models and Hypothesis. Vol. 1981. Heidelberg: Springer Verlag; 1981. pp. 1-9

[38] Ziganshin AU, Khairullin AE, Zobov VV, et al. The effects of ATP and adenosine on contraction amplitude of rat soleus muscle at different temperatures. Muscle & Nerve. 2017;**55**:417-423

[39] James RS. A review of the thermal sensitivity of the mechanics of vertebrate skeletal muscle. Journal of Comparative Physiology. B. 2013;**183**:723-733

[40] Ducharme MB, Van Helder WP, Radomski MW. Tissue temperature profile in the human forearm during thermal stress at thermal stability. Journal of Applied Physiology. 1991;**71**:1973-1978

[41] Ranatunga KW. Temperature dependence of mechanical power output in mammalian skeletal muscle. Experimental Physiology. 1998;**83**:371-376

[42] Bennett AF. Thermal dependence of muscle function. The American Journal of Physiology. 1984;**247**:R217-R229

[43] Rall JA, Woledge RC. Influence of temperature on mechanics and energetics of muscle contraction. The American Journal of Physiology. 1990;**259**:R197-R203

[44] Saltin B, Gagge AP, Stolwijk JA. Muscle temperature during submaximal exercise in man. Journal of Applied Physiology. 1968;**25**:679-688

[45] Kenny GP, Reardon FD, Zaleski W, et al. Muscle temperature of transients before, during, and after exercise measured using an intramuscular multisensor probe. Journal of Applied Physiology. 2003;**94**:2350-2357

[46] Yaicharoen P, Wallman K, Morton A, Bishop D. The effect of warm-up on intermittent sprint performance and selected thermoregulatory parameters. Journal of Science and Medicine in Sport. 2012;**15**:451-456

[47] Rowlerson AM, Spurway NC. Histochemical and immunohistochemical properties of skeletal muscle fibers from Rana and Xenopus. The Histochemical Journal. 1998;**20**:657-673

[48] Grishin SN, Ziganshin AU. Synaptic organization of tonic motor units in vertebrates. Biochemistry (Moscow), Supplement Series A: Membrane and Cell Biology. 2015;**9**:13-20

[49] Angilletta MJ. Thermal adaptation. In: A Theoretical and Empirical Synthesis. Oxford: Oxford University Press; 2009. 302 p

[50] Racinais S, Oksa J. Temperature and neuromuscular function. Scandinavian Journal of Medicine & Science in Sports. 2010;**20**:1-18

[51] Syme DA. Functional properties of skeletal muscle. In: Shadwick RE, Lauder GV, editors. Fish Physiology: Fish Biomechanics. Vol. 23. Amsterdam: Elsevier Academic Press; 2006. pp. 179-240

[52] Katz B. Neural transmitter release: From quantal secretion to exocytosis and beyond. Journal of Neurocytology. 2003;**32**:437-446

[53] Vyskocil F, Illes P. Non-quantal release of transmitter at mouse neuromuscular junction and its dependence on the activity of Na^+-K^+ ATP-ase. Pflügers Archiv. 1977;**370**:295-297

[54] Malomuzh AI, Nikol'skii EE. Non-quantum release of the mediator: Myth or reality? Uspekhi fiziologicheskikh nauk (Rus). 2010;**41**:27-43

[55] Vyskočil F, Malomouzh AI, Nikolsky EE. Non-quantal acetylcholine release at the neuromuscular junction. Physiological Research. 2009;**58**:763-784

[56] Nikol'skii EE, Voronin VA. Temperature dependence of the processes of spontaneous quantum and non-quantum release of the mediator from the motor nerve endings of the mouse. Neurophysiology. 1986;**18**:361-367

[57] Lupa MT, Tabti N, Thesleff S, et al. The nature and origin of the calcium-insensitive miniature end-plate potentials at rodent neuromuscular junctions. The Journal of Physiology. 1986;**381**:607-618

[58] Giniatullin RA, Khazipov RN, Oranska TI, et al. The effect of non-quantal acetylcholine release on quantal miniature currents at mouse diaphragm. The Journal of Physiology. 1993;**466**:105-114

[59] Glavinović MI. Voltage clamping of unparalysed cut rat diaphragm for study of transmitter release. The Journal of Physiology. 1979;**290**:467-480

[60] Datyner NB, Gage PW. Phasic secretion of acetylcholine at a mammalian neuromuscular junction. The Journal of Physiology. 1980;**303**:299-314

[61] Moyer M, van Lunteren E. Effect of temperature on endplate potential rundown and recovery in rat diaphragm. Journal of Neurophysiology. 2001;**85**:2070-2075

[62] Barrnett RJ. The fine structural localization of acetylcholinesterase at the myoneural junction. The Journal of Cell Biology. 1962;**12**:247-262

[63] Foldes FF, Kuze S, Vizi ES, Deery A. The influence of temperature on neuromuscular performance. Journal of Neural Transmission. 1978;**43**:27-45

[64] Kordas M. An attempt at an analysis of the factors determining the time course of the end-plate current. II. Temperature. Journal of Physiology. 1972;**224**:333-348

[65] Nikolsky EE, Strunsky EG, Vyskocil F. Temperature dependence of the carbachol-induced modulation of the miniature end-plate potential frequency in rats. Brain Research. 1991;**560**:354-356

[66] Asmussen G, Beckers-Bleukx G, Maréchal G. The force-velocity ratio of the rabbit inferior oblique muscle; influence of temperature. Pflügers Archiv. 1994;**426**:542-547

[67] Asmussen G, Maréchal G. Maximal shortening velocities, isomyosins and fibre types in soleus muscle of mice, rats and guinea-pigs. The Journal of Physiology. 1989;**416**:245-254

[68] Barclay CJ, Constable JK, Gibbs CL. Energetics of fast- and slow-twitch muscles of the mouse. The Journal of Physiology. 1993;**472**:61-80

[69] Leijendekker WJ, Elzinga G. Metabolic recovery of mouse extensor digitorum longus and soleus muscle. Pflügers Archiv. 1990;**416**:22-27

[70] Petrofsky JS, Lind AR. The influence of temperature on the isometric characteristics of fast and slow muscle in the cat. Pflügers Archiv. 1981;**389**:149-154

[71] Kössler F, Küchler G. Contractile properties of fast and slow twitch muscles of the rat at temperatures between 6 and 42°C. Biomedica Biochimica Acta. 1987;**46**:815-822

[72] Gilliver SF, Jones DA, Rittweger J, Degens H. Variation in the determinants of the power of chemically skinned type I rat soleus muscle fibres. Journal of Comparative Physiology. A, Neuroethology, Sensory, Neural, and Behavioral Physiology. 2011;**197**:311-319

[73] Binkhorst RA, Hoofd L, Vissers AC. Temperature and force-velocity relationship of human muscles. Journal of Applied Physiology: Respiratory, Environmental and Exercise Physiology. 1977;**42**:471-475

[74] Bottinelli R, Canepari M, Pellegrino MA, Reggiani C. Force-velocity properties of human skeletal muscle fibers: Myosin heavy chain isoform and temperature dependence. The Journal of Physiology. 1996;**495**:573-586

[75] Buller AJ, Kean CJ, Ranatunga KW, Smith JM. Temperature dependence of isometric contractions of cat fast and slow skeletal muscles. The Journal of Physiology. 1984;**355**:25-31

[76] Candau R, Iorga B, Travers F, et al. At the physiological temperatures the ATPase rates of shortening soleus and psoas myofibrils are similar. Biophysical Journal. 2003;**85**:3132-3141

[77] Gulati J. Force-velocity characteristics for calcium-activated mammalian slow-twitch and fast-twitch skeletal fibers from the guinea pig. Proceedings of the National Academy of Sciences of the United States of America. 1975;**73**:4693-4697

[78] Claflin DR, Faulkner JA. The force-velocity relationship at high shortening velocities in the soleus muscle of the rat. The Journal of Physiology. 1989;**411**:627-637

[79] Goldman YE, McCray JA, Ranatunga KW. Transient tension changes initiated by laser temperature in pbs. Muscle fibers. The Journal of Physiology. 1987;**392**:71-95

[80] Kochubei PV, SYu B. Comparison of the strength and speed of shortening of the fibers of fast and slow skeletal muscles of the crucible at different temperatures. Biophysics (Rus). 2014;**59**:967-972

[81] Rossi R, Maffei M, Bottinelli R, Canepari M. The temperature dependence of the speed of actin filaments propelled by slow and fast skeletal myosin isoforms. Journal of Applied Physiology. 2005;**99**:2239-2245

[82] Mukhamedyarov MA, Grishin SN, Zefirov AL, Palotas A. The mechanisms of the multi-component paired-pulse facilitation of the neurotransmitter release at the frog neuromuscular junction. Pflügers Archiv. 2009;**458**:563-570

[83] Grishin SN. Transmembrane calcium current: Mechanism,

registration procedures, Ca^{2+}-mediated modulators of synaptic transmission. Biochemistry (Moscow), Supplement Series A: Membrane and Cell Biology. 2014;**8**:213-224

[84] Grishin SN. Neuromuscular transmission in Ca^{2+}-free extracellular solution. Biochemistry (Moscow), Supplement Series A: Membrane and Cell Biology. 2016;**10**:99-108

[85] Rich TL, Langer GA. Effects of calcium-free perfusion on excitation-contraction in the heart and skeletal muscle. Recent Advances in Studies on Cardiac Structure and Metabolism. 1975;**9**:95-100

[86] Stephenson DG, Williams DA. Calcium-activated force responses in fast- and slow-twitch skinned muscle fibers of the rat at different temperatures. The Journal of Physiology. 1981;**317**:281-302

[87] Rutkove SB, Kothari MJ, Shefner JM. Nerve, muscle, and neuromuscular junction electrophysiology at high temperature. Muscle & Nerve. 1997;**20**:431-436

[88] Ferretti G. Cold and muscle performance. International Journal of Sports Medicine. 1992;**13**:S185-S187

[89] Boutilier RG, St-Pierre J. Adaptive plasticity of skeletal muscle energetics in hibernating frogs: Mitochondrial proton leak during metabolic depression. The Journal of Experimental Biology. 2002;**205**:2287-2296

[90] Mantovani M, Heglund NC, Cavagna GA. Energy transfer during stress relaxation of contracting frog muscle fibers. The Journal of Physiology. 2001;**537**:923-939

[91] Nyitrai M, Rossi R, Adamek N, et al. What are the limits of fast-skeletal muscle contraction in mammals? Journal of Molecular Biology. 2006;**355**:432-442

[92] Ziganshin BA, Elefteriades JA. Deep hypothermic circulatory arrest. Annals of Cardiothoracic Surgery. 2013;**2**:303-315

[93] Ziganshin BA, Rajbanshi BG, Tranquilli M, et al. Straight deep hypothermic circulatory arrest for cerebral protection during aortic arch surgery: Safe and effective. The Journal of Thoracic and Cardiovascular Surgery. 2014;**148**:888-898

[94] Andrews PJ, Sinclair HL, Rodriguez A, et al. Hypothermia for intracranial hypertension after traumatic brain injury. The New England Journal of Medicine. 2015;**373**:2403-2412

[95] Nielsen N, Wetterslev J, Cronberg T, et al. Targeted temperature management at 33°C versus 36°C after cardiac arrest. The New England Journal of Medicine. 2013;**369**:2197-2206

[96] Moler FW, Silverstein FS, Holubkov R, et al. Therapeutic hypothermia after out-of-hospital cardiac arrest in children. The New England Journal of Medicine. 2015;**372**:1898-1908

Chapter 3

Purinergic Signaling: A New Regulator of Ovarian Function

Mauricio Díaz-Muñoz, Anaí Campos-Contreras, Patricia Juárez-Mercado, Erandi Velázquez-Miranda and Francisco G. Vázquez-Cuevas

Abstract

Purinergic signaling is a sophisticated system of elements in which ATP and related molecules function as intercellular messengers. When ATP is released into the extracellular space, it activates specific receptors that belong to the P2 family. In parallel, ectonucleotidases transform ATP in its dephosphorylated metabolites including adenosine, which stimulates P1 receptors. The activity of both receptors influences various cellular processes. Moreover, metabolic conditions are concatenated with purine signaling to conform a dynamic and continuous informational network. The role of purinergic signaling in ovarian cells has been investigated, for instance, it is known that cells conforming the follicle express functional receptors that modulate basic cellular process such as proliferation, induction of apoptotic cell death, and steroidogenesis. In this chapter, we review contemporary information on purinergic action in ovarian cell physiology and state its relevance in this field.

Keywords: purinergic signaling, ovary, granulosa, theca, OSE

1. Introduction

The ovary is a complex cell system where folliculogenesis and steroid hormone synthesis take place. These processes involve dynamic changes in the cellular populations of this tissue and highly precise mechanisms of regulation. Folliculogenesis requires a complex coordination of three stages: *recruitment*, which is the initial growth of a group of follicles from the reserve pool; *selection*, during which a subgroup from this pool of recruited follicles survives and grows, while the remainders suffer atresia; and *dominance*, in which the follicles that will be ovulated reach the preovulatory stage and increase their size, while the subordinate follicles arrest their growth. It is well known that the initial stages of folliculogenesis are independent of gonadotropins, whereas the advanced stages depend on these hormones. At the same time, *selection* and *dominance* stages implicate the elimination of subordinated follicles by apoptosis; thus, a coordinated set of events directs growth and surveillance of some follicles and disappearance of others with extreme precision [1].

Folliculogenesis involves constant rearrangement in ovarian cellular architecture. A primordial follicle is formed by an oocyte surrounded by a layer of squamous epithelial cells and is arrested in the diplotene phase of meiosis I. When it is recruited and enters the growing stage, granulosa cells change their shape from flat to cuboid, starting their proliferation; this entity is known as *primary follicle*. The granulosa layer continues proliferating to adopt a stratified epithelium; then, the theca layer surrounding the follicle emerges. This layer is innervated and vascularized conforming a *secondary* follicle. Inside the follicle, three to four cavities filled with fluid are formed; the cavities fuse and form the antrum. This stage is known as antral follicle. Antrum formation implicates the emergence of two granulosa populations: cumulus cells in contact with the oocyte and mural cells attached to the follicular theca. These follicles grow until reaching the preovulatory stage [1]. Changes in follicle development are orchestrated by diverse cellular messengers (reviewed in [2]).

Moreover, synthesis and secretion of ovarian steroid hormones are coordinated by the somatic components of the follicle; thus, cholesterol and acetate uptake from the blood by the theca layer allows the synthesis of androgens that are aromatized into estrogens in the granulosa cells [3]. This process is finely regulated by the coordinated action of intraovarian and endocrine components.

All these dynamic processes are organized at distinct levels by endocrine, nervous, and autocrine-paracrine mechanisms acting with high systematic precision. Recent findings suggest that purinergic signaling participates in the control of cellular process in the follicle cells, making it a new player in the network of signals regulating the cellular biology of ovarian physiology. Given that follicle cell types have defined roles and are typical for each follicular growth stage, receptor expression in each of these cell types suggests a potential physiological role; thus, we organized the information of this chapter to cover the role played by purinergic receptors in ovarian physiology.

2. Purinergic system

From a chemical perspective, purines are defined as nitrogen-containing heterocyclic aromatic compounds formed by the fusion of a pyrimidine and an imidazole ring (in total, five carbon and four nitrogen atoms). The purine molecule can be associated with an amino group, such as adenine (6-amino purine), or with amino and keto groups, such as guanine (2-amino-6-oxy purine). The existence of functional groups that are weakly acidic next to a system of conjugated dienes favors the relocation of a proton along the molecule, allowing the formation of constitutional isomers called tautomers. The principal tautomeric equilibriums in purine molecules are amino (predominant)-imino forms and lactam (predominant)-lactim forms [4].

Purines are part of important informational polymers, such as nucleic acids (DNA and RNA), and of various biomolecules with metabolic and functional significance such as bioenergetic factors (ATP and GTP), redox coenzymes (NAD(P)H/NAD(P)$^+$, FMNH$_2$/FMN$^+$, and FADH$_2$/FAD$^+$), biochemical crossroad metabolites (coenzyme A), and signaling molecules (cyclic AMP, cyclic ADP-ribose, nicotinic acid adenine dinucleotide phosphate [NAADP]) and ligands for purinergic receptors. Caffeine and theobromine are naturally occurring purines in plants [5]. In this section, we will refer exclusively to ATP, adenosine (ADO), and other related molecules as factors in cellular communication and coordinators of metabolic networks (**Figure 1**).

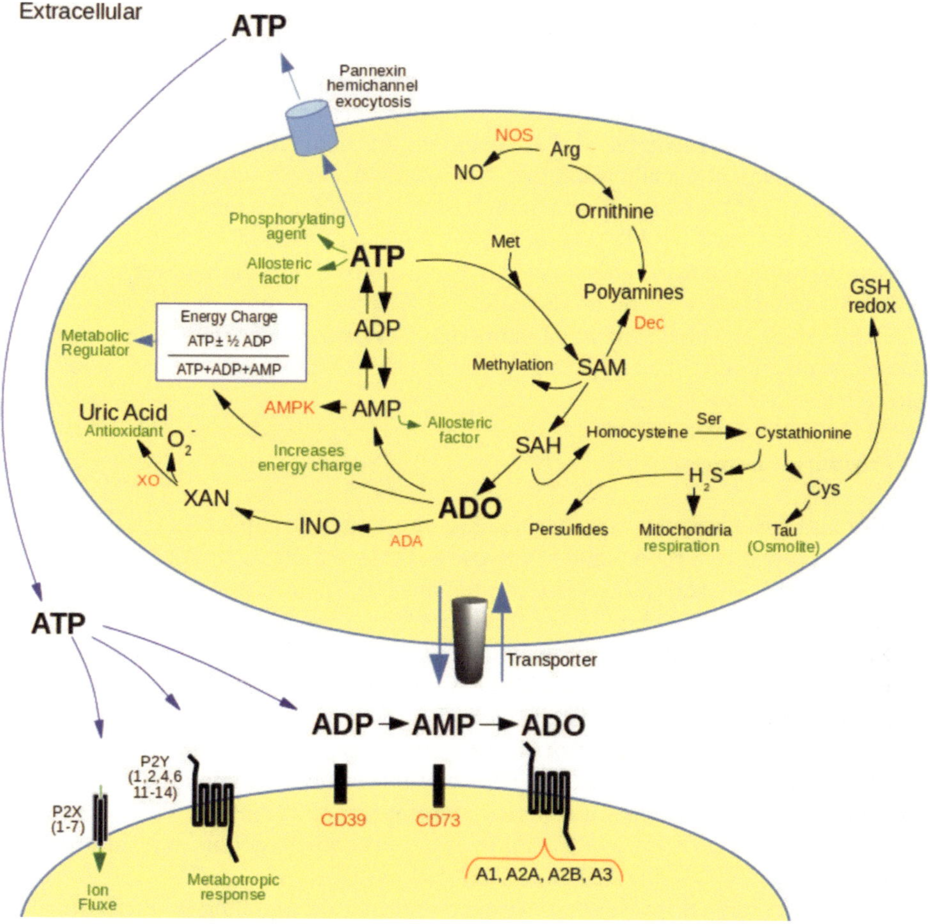

Figure 1.
Purines in cellular signaling and metabolic networks. The figure depicts two cells in communication using purines as signaling molecules. Released ATP is recognized by two types of receptors (P2X, ionotropic, and P2Y, metabotropic). ATP can be hydrolyzed by different ectonucleotidases (such as CD39 and CD73) to be transformed in less phosphorylated nucleotides (ADP and AMP) and eventually to the nucleoside adenosine (ADO). The receptors that mediate ADO's actions are all metabotropic. Thus, purinergic signaling is the result of the types of receptors present in the cellular system, as well as the variable proportion of adenine nucleotides and ADO that result from the enzymatic activities of the expressed ectonucleotidases. There are membrane transporters with the capacity to transfer ADO to the cellular inside. Within the cell, ADO can be transformed in multiple metabolites and act as modulator of strategic metabolic pathways. ADO is catabolized by adenosine deaminase (ADA) resulting in a set of purines that culminates with uric acid (a mild antioxidant) with xanthine as an intermediary with the capability to generate the anion superoxide (O_2^-). By the action of adenosine kinase, ADO can turn into a nucleotide (AMP) that is a modulator of the AMPK, which is an important enzyme to regulate energy homeostasis. AMP can be further phosphorylated to ADP and ATP. Eventually, ATP can go to the extracellular space to fulfill its messenger role by exocytosis or by specialized channels (pannexin and connexin hemichannels). The proportion of adenine nucleotides (energy charge) is an important modulator of the equilibrium between catabolic and anabolic reactions. It has been reported that ADO is able to increase the cellular energy charge in the liver. When ADO binds homocysteine to form S-adenosylhomocysteine (SAH), it can also influence the transmethylation and transsulfuration pathways. SAH is usually formed when the methylating agent S-adenosylmethionine (SAM) transfers a methyl moiety to a given substrate. Interestingly, SAM can also participate in the synthesis of polyamines by transferring an aminopropyl group from decarboxylated SAM to putrescine. Homocysteine is joined to serine to form the intermediary cystathionine, which is transformed into cysteine and the intracellular messenger H_2S. Meanwhile, cysteine can be incorporated into the redox and antioxidant molecule glutathione (GSH) or can be converted into the osmolyte taurine. Hence, a condition in which ADO is increased beyond a threshold may accumulate SAH with the consequent interruption of the methylating reaction and the promotion of polyamine synthesis, as well as the reduction of homocysteine availability and the decrease in cystathionine and all its derivatives.

Adenine-related molecules are ubiquitously present in all living beings. ATP (a nucleotide) and ADO (a nucleoside) are easily interconverted by a set of three phosphorylation/dephosphorylation steps. Interestingly, this purine conversion involves different mechanisms when it takes places either inside or outside the cellular milieu. Intermediates of these reactions are AMP and ADP nucleotides, whereas some metabolically important ADO derivatives are uric acid, S-adenosyl methionine (SAM), and S-adenosylhomocysteine (SAH). Various physiopathological events modulated by ATP and/or ADO have been reported, including sleep, immunity, tumorigenesis, platelet aggregation, vasodilation, inflammatory and hypoxic responses, and antioxidant status [6]. Overall, adenine nucleotides and ADO are interconnected signaling factors that activate specific membrane receptors and act as metabolic regulators that coordinate the equilibrium between anabolic and catabolic reactions [7].

2.1 Purinergic communication

ATP and ADO are well-known signaling molecules. Both purines have the capacity to promote a set of cellular responses by acting through specific membrane receptors, either by activating ionic conductance or by promoting the formation of second messengers. Two families of receptors for purine ligands have been characterized: (1) P1 or ADORA receptors, which are G-coupled metabotropic adenosine receptors and are classified as A1 and A3 (associated with adenylate cyclase inhibition and formation of IP_3 and diacylglycerol) and as A2A and A2B (both activate adenylate cyclase), (2) P2 receptors, which recognize ATP as a principal ligand but also a variety of related compounds. P2 receptors are divided into ionotropic receptors (P2X1–7) and G-coupled metabotropic receptors (P2Y1, P2Y2, P2Y4, P2Y6, P2Y11–P2Y14). In addition to ATP, P2Y receptors can be recognized by alternative ligands such as ADP, UTP, UDP, and UDP glucose [8].

Unlike other signal transduction systems, purinergic signaling involves the actions of two interconvertible ligands, ATP and ADO, within the same system. Interestingly, these ligands exert complementary or antagonistic actions on each other [9]. For example, epithelial-mesenchymal transition (EMT), a cellular plasticity process important in phenotypic programming of metastatic tumors, can be upregulated by ATP in SKOV-3 cells (from ovarian carcinoma) but downregulated by ADO [10]. This circumstance indicates that enzymes (ectonucleotidases) allowing the sequential conversion from ATP to ADO are a potential regulatory node that controls diverse cellular and physiological responses.

Four families of extracellular enzymes that transform purine nucleotides into ADO and phosphate are known: (1) ENTPD/CD39 (ectonucleoside triphosphate diphosphohydrolase) forms AMP from ATP/ADP; (2) ENPP (ectonucleotide pyrophosphatase) forms AMP from ATP/ADP/ADP-ribose; (3) alkaline phosphatase hydrolyzes different nucleotides to be transformed into ADO; and (4) NT5E (ecto-5′-nucleotidase) forms ADO from AMP [11]. It is expected that the combined and sequential activities of these enzymes in a given cell system result in a highly variable and potentially fine-tuned proportion of adenine nucleotides (ATP, ADP, and AMP) and ADO. Therefore, the physiological equilibrium between ATPergic and adenosinergic transmissions should be considered as a unique and emergent property of each cell system.

2.2 Purine-related metabolites

Adenine purines are key elements in metabolic network control. The proportion of adenine nucleotides dictates the direction of anabolic and catabolic processes by means of the regulatory parameter known as energy charge (ATP + ½ADP/ ATP + ADP + AMP). In this context, ATP and AMP act as allosteric factors for a variety of enzymes that are important in the bioenergetic status of the cell: phosphofructokinase 1 (glycolysis), aspartate carbamoyltransferase (pyrimidine synthesis), and glycogen phosphorylase (glycogenolysis). Interestingly, ADO treatment is one of the few cases that can upregulate the energy charge in vivo [12]. Another example of how the AMP/ATP ratio influences the metabolic networks is AMP kinase (AMPK) activation. AMPK is a strategic kinase that modulates the fasting response by phosphorylating and activating key catabolic enzymes [13].

ADO is a crossroad metabolite; it can be turned into nucleotides (first in AMP by adenosine kinase), or it can originate active catabolites such as xanthine (a source of superoxide) and uric acid (a terminal metabolite and mild antioxidant). The production of ADO in the liver and its transport through the blood are controlled by the circadian timing system; eventually, these mechanisms allow the 24-h rhythmic presence of purine rings in the nervous system, which are necessary for the onset of sleeping [14]. ADO has also been used as a hepatoprotective and antitumoral agent [15].

ADO also plays a role in the transmethylation and transsulfuration pathways, with SAM as a central metabolite for both. Initially, ATP activates methionine, which in turn generates SAM, the main cellular methylating agent. Some principal methylated molecules are phospholipids (phosphatidylethanolamine turns into phosphatidylcholine), catecholamines (adrenaline turns into noradrenaline and serotonin into melatonin), and nucleic acids (during epigenesis and RNA processing). SAM is transformed into SAH, which is hydrolyzed into ADO and homocysteine. In this metabolic step, ADO levels can modulate the methylation process, as high ADO favors SAH synthesis, thus blocking methyl donation. SAM is also a substrate for polyamine synthesis. In the presence of serine, homocysteine is converted into cystathionine. Within the mitochondria, this intermediate is the precursor of both the gasotransmitter H_2S and the principal antioxidant agent, glutathione [16].

2.3 Membrane transporters

Extracellular and intracellular purine dynamics are interconnected, mainly through specialized protein transporters that allow ATP and ADO to transit throughout the plasma membrane.

ATP exists at millimolar levels within the cell. It exits to the extracellular space, where it acts as a cellular messenger through two main paths: (1) exocytic, which involves secretion of vesicles mainly derived from Golgi and secretory ATP-containing granules by means of Ca^{2+}-dependent membrane depolarization and (2) conductive, in which the ATP efflux is carried out by different ion channels: hexamers of connexin subunits, assemblies of pannexin subunits, volume-regulated anion channels, and maxi-anion channels [17]. As mentioned, ATP can lose its phosphates in the extracellular space and transform into ADO. In turn, ADO can return to the intracellular milieu by the action of two types of carriers; one is driven by a facilitated diffusion event (sensitive to dipyridamole), and the other mobilizes ADO by an active process regulated by the Na^+-transmembrane gradient [18]. Overall, a net flux of purines can be visualized: first, purine rings exiting as ATP and eventually purines returning to the cell interior as ADO. Indeed, this cycle involves the net efflux of phosphate as well as the net conversion of intracellular ATP into ADO.

2.4 Integrative considerations

The transformation among purine molecules inside and outside the cell has the potential of intricately regulating both purinergic signaling and metabolic control.

To propose a model of this interaction, the following considerations are necessary: (1) at least three compartments should be taken in account—extracellular, cytoplasmic, and mitochondrial. However, if there is segregation of receptor populations in different membrane domains (e.g., signalosomes), the extracellular compartment could be more complex. (2) It is important to know the principal metabolite and intermediate levels in the process; indeed, the concentrations of these factors are expected to fluctuate, but knowing average levels is necessary. For example, ATP is at [mM] in the intracellular milieu, but when it is released in the pericellular space, it changes from [nM] to [mM] [19]. (3) It is important to know which receptors are present in a given cell system, as well as their physical constants (Kd and Bmax). (4) It is convenient to have a clear idea of the activities and regulation of all ectonucleotidases and their corresponding carriers. (5) In the same context, determining the presence of purine metabolizing enzymes and kinetic constants (Km and Vmax) is required. (6) It is necessary to know the conformational status of all proteins involved in the purine cycle in order to detect potential allosteric modulation.

3. Purinergic signaling in the ovary

3.1 Granulosa and luteal cells

Early studies analyzed the effects of ADO over the gonadotropin-induced cAMP production in granulosa and luteal cells. In rat and human granulosa cells, ADO incremented the accumulation of cAMP in response to follicle-stimulating hormone (FSH); similar effects were observed when human luteal cells were stimulated with human chorionic gonadotropin (hCG) or luteinizing hormone (LH), revealing the possibility that ADO is a gonadotropic modulator [20–23]. According to the dual role of ADO in metabolism and cell signaling, it was originally proposed that ADO effects were mediated by two mechanisms acting synergically: ADO translocation to the cytoplasm where the nucleoside can sustain the increment of cytosolic ATP and extracellular activation of specific membrane receptors [20].

In primary cultures of rat granulosa cells, incubation with ADO incremented intracellular ATP; dipyridamole blocked this effect, indicating ADO uptake through specific transporters [22]. The regulation of adenylate cyclase (AC) activity by ADO analogues was investigated in membranes isolated from the whole ovaries in follicular growth. It was observed that adenosinergic agonists incremented adenylate cyclase (AC) activity. Pharmacological approaches suggested that the effect of ADO analogues was mediated by the A2A receptor because it was strongly promoted by 5'-N-ethylcarboxamidoadenosine (NECA) and antagonized by 8-phenyltheophylline (8-PTH) [24]. Similar findings were observed in membranes from luteinized ovaries from superovulated rats induced by injection with pregnant mare serum gonadotropin (PMSG) and in homogenates of isolated cells from luteal bodies; moreover, it was shown that ADO analogues incremented progesterone synthesis. In these preparations, pharmacological evidence also indicated the participation of A2A receptor in the adenosinergic induction of cAMP accumulation, suggesting that ADO is a paracrine regulator of the luteal body's endocrine activity [25].

In addition, it has been shown that in ovine luteal cells, ADO increases the effect on luteotropins [hCG] as well as prostaglandins (E1 and E2) over progesterone production and inhibits the antigonadotropic and luteolytic effect of prostaglandin F2α (PGF2α) [20, 26]. Purinergic responses in ovarian cells were described before cloning P2 receptors. The effect of adenine nucleotides on cytosolic concentration of Ca^{2+} ($[Ca^{2+}]_i$) was first investigated in human luteal cells and in porcine granulosa cells; it was observed that nucleotides elicited an increment of $[Ca^{2+}]_i$ in both cell types. Moreover, in luteal cells, these compounds also induced an increase in progesterone and estradiol secretion [27]. Consistently, in granulosa from hen preovulatory follicles, ATP, and other adenine nucleotides also induced an increment of $[Ca^{2+}]_i$; the pharmacological characterization of this response revealed that it was mediated by intracellular Ca^{2+} release and dihydropyridine-insensitive Ca^{2+} channels according to P2Y receptor activation [28]. In human granulosa cells, it was determined that ATP responses were dependent on Ca^{2+} released from intracellular stores [29].

The molecular description of the P2Y2 receptor (then named P2U because of its sensitivity to UTP) in human granulosa-luteal cells (GLC) was made by Leung's group in British Columbia [30]. They detected the transcript of P2U receptor by Northern blot and reported the elevation of cAMP promoted by hCG. The stimulation of these cells with UTP/ATP induced an increment of $[Ca^{2+}]_i$ associated with phospholipase C (PLC) activation. Downstream of this pathway, protein kinase C (PKC) was activated, and it negatively modulated the P2Y2-dependent $[Ca^{2+}]_i$ response [31]. The molecular machinery and mechanism involved in the purine-induced increment of $[Ca^{2+}]_i$ have been studied. ATP-induced Ca^{2+} release mediated by activation of PLC and inositol triphosphate (IP_3) production. Indeed, IP_3 and ryanodine receptor (RyR) expression in GLC has been demonstrated. The increment of ATP-induced $[Ca^{2+}]_i$ was modulated by substances interfering with the activity of both RyR and IP_3R, revealing an interplay between both receptors to amplify the purinergic $[Ca^{2+}]_i$ signal [32]. Moreover, it was described that GLC expresses three isoforms of IP_3R, RyR, and thapsigargin-sensitive Ca^{2+}-ATPase (SERCA) [33].

Electrophysiological and Ca^{2+} imaging studies in the GFSHR-17 cell line from granulosa determined that P2Y2 and P2Y4 receptor stimulation induces Ca^{2+} mobilization and hyperpolarization. Both responses were sensitive to the PLC inhibitor U73122 and to the IP_3R antagonist 2-aminoethyl diphenyl; hyperpolarization was mediated by Cl^- channels, probably dependent on intracellular Ca^{2+} [34].

Moreover, it was shown that the P2Y2 (P2U) activation in human GLC induced a decrease in the LH-dependent cAMP levels; this antigonadotropic effect was mediated by PKCα activity [35]. In parallel, it was shown that P2Y2 stimulation also activated extracellular mitogen-regulated kinases (ERK) through a Gαq-dependent pathway; ERK activity was responsible for inhibiting LH-dependent production of progesterone induced by P2Y2 activation [36]. Further experiments revealed that phospho-ERK translocates to the nucleus and regulates cell proliferation by early growth-1 (egr-1) and c-raf-1 responses [37]. These data clearly show that the purinergic response mediated by P2Y2 in GLC can be an important modulator of gonadotropic actions and granulosa cell physiology.

Expression of the P2Y6 receptor was observed in murine GLC. Its stimulation with the selective agonist UDP incremented cell viability and progesterone but did not affect estradiol production. This effect on steroidogenesis was concomitant with a negative regulation of enzymes corresponding to Δ4 steroidogenic pathway, CYP11A, 3β-HSD, and StAR. The effects were blocked by the antagonist MRS2578 before UTP stimulus [38]. These results suggested that purines acting through P2Y6 regulate luteal body viability and steroidogenic function.

In human GLC, extracellular ATP promoted apoptosis by activating P2 receptors. These receptors elevated $[Ca^{2+}]_i$, which in turn activated Ca^{2+}-dependent K^+ channels, leading to membrane depolarization [39]; however, the specific receptor involved in this response was not identified. These observations were confirmed by a later study that demonstrated the participation of apoptotic marker caspase-3

[40]. This effect was apparently not consistent with previous observations regarding ATP actions in GLC, but it could be explained by the differential sensitivity of distinct P2 receptors [41]. It is possible that extracellular concentration of ATP and the expression of P2 receptor determine the specific effect of purinergic stimulation.

Recently it was described that the P2X7 receptor is expressed in mouse luteal cells. Its activation with ATP or BzATP induced an antiproliferative effect by regulating the expression of cyclin D2 and cyclin E2, as well as the phosphorylation of mitogen-activated protein kinase p38 [42]. The result suggests a role for the P2X7 receptor in luteal body function.

The purinergic system is well represented in GLC. A set of purinergic receptors can modulate basic cellular processes such as proliferation, apoptosis, and steroidogenesis. Growing evidence indicates that purines are important regulators of GLC, but further studies are necessary to reinforce their role in ovarian physiology.

3.2 Theca cells

The theca cell layer is an enclosure of cells that surrounds the oocyte during folliculogenesis. It is crucial for maintaining the structural integrity of the follicle as well as for regulating nutrient influx to the avascular GCL [43, 44]. Theca is also the site for the synthesis of steroid hormones, specifically androgens (testosterone and dihydrotestosterone), from acetate or cholesterol into estrogens by granulosa cells in an LH-dependent manner [3]. In addition, theca is the only component of the follicle that is innervated by sympathetic and parasympathetic nervous systems, implicating that this layer functions as a complex integrator of endocrine and neural information [45].

When a primary follicle has one or two layers of granulosa cells, an outer granulosa cell layer differentiates into theca cells and, together with recruited theca precursor cells from the stroma, forms the theca cell layer surrounding the oocyte [44, 46]. Some pathophysiological reproduction-related conditions such as infertility or polycystic ovarian syndrome are often the result of dysfunctional activity of theca cells during ovulation and follicle development [47, 48].

The theca cell layer contacts the rich microvasculature system surrounding each follicle and integrates signals from autonomic innervation [49]. It has been demonstrated that ATP can be co-released with noradrenaline from terminals of the peripheral nervous system [50] and as a result of mechanical stress and changes in cell volume in the oocyte [51]; thus, ATP is a relevant modulator of cellular communication between the theca cell layer and surrounding oocyte cells.

Purinergic signaling has been described in female reproductive organs, and evidence has shown that ATP in the extracellular space participates in the physiological regulation of the ovary [52]. The first characterization of purinergic signaling in theca cells showed the functional expression, and activation of P2X7 receptors induced cell death, an important mechanism for the onset and physiological progression of follicle atresia [53]. P2X7 receptors have also been associated with the inhibition of luteal cell survival and proliferation, pointedly in small luteal cells, which have been suggested as theca-derived luteal cells [42].

On the other hand, there is also evidence of the expression of uridine triphosphate (UTP)-sensitive P2Y receptors in theca cells P2Y2 and P2Y6, but not P2Y4 [54]. In this system, stimulation of the expressed P2Y receptors with UTP in cultured theca cells induces the activation of mitogenic-signaling pathways that promote cell proliferation [54]. This finding is a relevant pathophysiological indication, since a slow but maintained proliferation takes place in polycystic ovarian syndrome [47].

Furthermore, an interaction between adenosine receptor A2 and P2Y receptors has been described in theca of *Xenopus* ovarian follicles. The authors suggested that this association took place when both the epithelial and theca cell layers of the oocyte were intact [55].

Collectively, these findings suggest that a tight regulation of purinergic expression and signaling must be in place for the theca cell layer to function properly and communicate with neighboring cells.

3.3 Cumulus complex

The organized structure of the cumulus-enclosed oocyte (CEO) complex corresponds to a specialized GLC surrounding the oocyte. Cumulus cells secrete factors to regulate oocyte maturation and maintain meiotic arrest [56]. It was reported that porcine and murine follicular fluid contains purine compounds that presumably participate in CEO functions, suggesting that it could be an important signal to trigger physiological events [57, 56]. Until recently, purinergic receptors were identified and characterized in CEO, indicating that purinergic signaling participates in CEO physiology [58, 59].

When Eppig et al. discovered that the main components of follicular fluid were nucleotide-derived metabolites [60] and established a relationship with follicle maturation [56, 57], they hypothesized that the local purinergic metabolism in the ovarian fluid can be involved in oocyte maturation or may participate in other aspects of follicular functions.

In these studies, the concentration of nucleotide compounds in murine follicular fluid was determined using high-performance liquid chromatography (HPLC). They identified two purine compounds: hypoxanthine and ADO, with concentrations in ranges of 2–4 and 0.3–0.7 mM, respectively. They also showed that these purines affected the CEO by maintaining the meiotic arrest [56].

Eppig et al. also analyzed the same compounds in porcine follicular fluid; they identified that hypoxanthine at 1.4 mM is the major inhibitory component producing a transient inhibition [61]. This observation contrasted with that of other laboratories which had failed to detect inhibitory activity in follicular fluid.

On the other hand, the cellular effects of purinergic ligands were studied by Ca^{2+} imaging and electrophysiological approaches. In 2002, two reports elucidated which purinergic receptor was expressed in CEO cells. The first report of P2Y expression in CEO began with an interesting observation that ATP could stimulate an intracellular Ca^{+2} transient. Experiments using the CEO complex and applying ATP or UTP to the extracellular solution induced a wave of Ca^{+2} mobilization from cumulus cells to the oocyte through gap junctions, suggesting that ATP was involved in oocyte maturation; moreover, they showed that the response involved the P2Y2 receptor. Since gonadotropin hormones, follicle-stimulating hormone (FSH), and luteinizing hormone (LH) had no effect on Ca^{+2} changes [59], the authors concluded that ATP was the specific messenger to mediate calcium signals involved in oocyte maturation in the CEO complex.

In another report, the responses generated by a putative purinergic receptor expressed in the CEO were identified and characterized. Employing the voltage clamp technique with two electrodes, the authors observed depolarization responses when extracellular ATP was applied. RT-PCR analysis revealed a product correspondent to the P2Y2 receptor, suggesting that calcium mobilization is dependent on this receptor. A detailed description of distinct currents generated from several ionic channels, such as Ca^{+2}-dependent Cl^- current, voltage-dependent K^+ currents, and a cationic current mainly driven by Na^+, was provided. The authors concluded that both purinergic receptors and ionic channels were located in CEO cells that transmitted their electrical signals to the oocyte via gap junctions [58]. These data support the idea that P2Y2 is an important element in paracrine signaling in regulating CEO complex physiology. Future studies are required to determine the mechanisms involved in CEO functions and oocyte maturation by ATP stimulation.

3.4 Ovarian surface epithelium

Ovarian surface epithelium (OSE) is a monolayer surrounding the ovary. It is composed of a single flat layer of squamous-to-cuboidal epithelial cells featuring distinguished epithelial and mesenchymal markers. OSE is essential during ovulation to promote follicular rupture and release the oocyte [62] and for postovulatory repair of the ovary [63].

Initial and important studies were led by Nelly Auersperg to characterize and identify epithelial and mesenchymal markers, hormonal and growth factor receptors, and physiopathological role, with the idea that OSE is determinant in the onset of ovarian carcinoma [64]. During ovulation, the OSE is involved in three main phases: apex formation, rupture, and repair [65].

The initial phase starts with the actions of luteinizing hormone (LH) and triggers apex formation at the rupture site of the ovarian surface [66]. In the second phase, OSE cells initiate a lytic cascade [62, 67, 68], releasing proteolytic enzymes to degrade the basal lamina, the tunica albuginea, and ovarian cells of the mature follicle. The digested matrix, follicular wall disintegration, and peeling of OSE cells create a wound stigma that facilitates oocyte release. Finally, the repair phase consists in wound closure by postovulatory cell proliferation and migration [69].

Nevertheless, the signaling involved in these phases during ovulation in the OSE is unclear. Purinergic signaling was suggested as a part of intraovarian modulation due to a certain purinergic receptor expressed in the OSE committed in physiological processes. Recently, Vazquez-Cuevas et al. demonstrated the expression of ligand-activated ion channel P2X7 in OSE. In primary cultures of mouse OSE, they observed that BzATP induced a non-desensitizing increment of $[Ca^{+2}]_I$, and this response was blocked with A438079, a selective antagonist of the P2X7 receptor. The functional role of P2X7 was investigated in situ by TUNEL assay. P2X7 stimulation with BzATP induced apoptosis in OSE cells and was differential throughout the oestrous cycle; DNA fragmentation was greater during proestrous [70]. These findings contribute to the idea that local factors, such as ATP, may participate in a proposed cyclic proliferation-death equilibrium of the OSE cell layer in the ovulatory process.

Understanding purinergic signaling and receptor expression in the OSE will help to decipher the mechanisms underlying ovary physiology and pathology. However, more studies need to contribute evidence related to homeostasis and postovulatory repair. Some studies regarding OSE-derived cancer cells will be discussed in the next section.

4. Purinergic signaling in ovarian cancer

Although the specific roles for purinergic signaling in ovarian physiology are not completely understood, significant advances have been made in deciphering the role of purines in cancer. Plenty of information supports that purinergic system elements have a role in cancer progression, and this implicates that they are potential therapeutic targets. Here we will address the relevance of the purinergic system in ovarian cancer (OC).

Ovarian cancer (OC) is considered the most lethal gynecological malignancy, as it is usually diagnosed by the time the tumor has spread to other regions [71]. OC is the seventh most common type of cancer in women, and patients have a low survival rate [72]. Early detection of the disease proves difficult due to unspecific symptoms such as abdominal pain and bloating, whereas advanced stages are confused with gastrointestinal illnesses [73]. Although OC can arise from any of the cells located in the ovary (germ, stroma, and the epithelium), it is acknowledged that almost 90% of OC is derived from the OSE. According to histological characteristics, epithelial ovarian cancer (EOC) is classified in serous, endometrioid, and clear cells and in mucinous carcinomas [74].

Ovarian tumors frequently show disseminated metastasis through ascites in the peritoneal cavity. OC cells are exfoliated from the primary tumor surface to the peritoneal fluid, where they survive as single cells or multicellular aggregates. These cells acquire resistance to anoikis, have stem cell properties, and are plastic in terms of switching between epithelial and mesenchymal phenotypes. In addition to cancer cells, the malignant ascite microenvironment has normal cell types, such as platelets, associated fibroblast, and immune cells, which support and assist cancer cells [75, 76]. This distinctive tumor microenvironment (TME) has paracrine and autocrine signals that support cancer cell proliferation, death evasion, dissemination, and invasion to peritoneal organs. Therefore, understanding cellular and molecular mechanisms that promote progression of the disease is very relevant.

Purinergic signaling has emerged as an important regulator of tumor growth [77]. The following facts support this assertion: (1) cancer cells increase in metabolic rate [78]; (2) ATP and ADO levels increase in the tumor interstitium [79, 80]; (3) purinergic receptors are expressed in tumor cells [77]; and (4) high CD73 ectonucleotidase expression is a prognosis factor for several types of cancer [81].

Early studies of purinergic signaling in ovarian carcinoma-derived cells (OCDC) evaluated the effects of ATP on $[Ca^{2+}]_i$ mobilization. It was demonstrated in OVCAR-3 and SKOV-3 cell lines that stimulation with ATP in a μM range induced a biphasic response that consisted in a rapid peak followed by a smaller and sustained plateau phase. In addition, chelation of Ca^{2+} abrogated the slower response induced by ATP, while the rapid response was maintained. Furthermore, low concentrations of ATP induced SKOV-3 and OVCAR-3 cell proliferation [82, 83]. These pioneer works indicated that OCA cells were responsive to ATP. Furthermore, the biphasic response induced by ATP suggested that these cells could express two types of purinergic receptors: channels operated by ligands and G protein-coupled receptors (GPCRs).

In other OCDC lines (EFO-21 and EFO-27), the presence of P2Y2 receptor was described, and ATP response was monitored through intracellular Ca^{2+} mobilization and phospholipase D (PLD) activation. In addition, ATP downregulated basal cell proliferation and the proliferation induced by fetal calf serum (FCS) [84]. However, in IOSE-29 (preneoplastic) and IOSE-29EC (neoplastic) cell lines, which also express P2Y2, ATP-stimulated cell growth through the MAPK/ERK kinase activation [85]. These findings showed that ATP effects differ across cell lines, which could be associated with the activated intracellular mechanism.

Regarding the expression of P2X receptors in OCDC lines, some of the first reports evaluated P2X7 expression in OC biopsies as in SKOV-3 and OVCAR-3 cells. P2X7 expression in human ovaries is confined to the OSE, whereas in EOC biopsies, its expression is wider and localized in transformed zones [86, 87]. Additionally, receptor functionality was evaluated in cell lines, and it was demonstrated that its stimulation induces ERK and AKT phosphorylation, whereas its inhibition reduces cell viability [87]. The latter result was surprising due to previous findings that associated P2X7 receptor activation with apoptosis.

An important feature of cancer cells is their ability to migrate and invade secondary organs. One process that allows OC cells to dissociate from primary tumors and survive in peritoneal fluid is the EMT, in which cells switch from an epithelial to a mesenchymal phenotype. Even though this process was first described in embryonic development, today its role in cancer is well accepted. The involvement of purines in EMT has recently been reviewed [9]. It has been proven that P2Y2 stimulation promotes SKOV-3 cell migration, and this effect is associated with epithelial growth factor receptor (EGFR) transactivation. Moreover, expression of EMT inductors such as SNAIL and TWIST was promoted in response to UTP; in addition, the intermediate filament vimentin was augmented by this pharmacological stimulus. Interestingly, addition of apyrase (Apy) to cell medium, with the aim of removing extracellular ATP, decreased cell migration and favored an epithelial-like phenotype due to the relocation of E-cadherin to SKOV-3 cellular junctions [88]. The authors concluded that products obtained by ATP hydrolysis (i.e., ADO) promoted an epithelial phenotype, while P2Y2 activation by ATP analogues promoted a mesenchymal one.

As previously mentioned, ATP and ADO concentrations increase in cancer. Although ATP is within a low nM range in the healthy interstitium, it increases to μM concentrations (~800 μM) near the tumor. This new evidence is relevant because extracellular ATP was monitored in vivo in tumors induced with OVCAR-3 cells in nude mice [80]. Given that extracellular ATP can be hydrolyzed by membrane ectonucleotidases (CD39 breaks ATP into ADP and AMP, and CD73 hydrolyzes AMP to ADO), it could be assumed that the increase in extracellular ATP is directly correlated with the increase in ADO. Even though extracellular ADO has not been measured in tumors in vivo, there is evidence that ADO levels increase in tumor microdialysates and are more abundant at the core of the tumor [79]. In vitro studies have also indicated that OCDC-derived cell lines release ATP to the cell medium [10, 89].

Correspondingly, ADO function has been evaluated in several types of cancer, and an immunosuppressive role has been proposed for this molecule through ADORA2A receptor activation. Specifically, in OCDC cell lines, it has been demonstrated that ADO inhibits antitumor activity of T and natural killer (NK) cells. Moreover, OC biopsies, OC-derived primary cultures, and cell lines express functional CD39 and CD73 [89]. Therefore, strategies such as CD39 and CD73 inhibition with the aim of reducing extracellular ADO concentrations have been performed, and an improved immune response was described using antibodies against these enzymes [89]. CD73 expression has also been associated with poor prognosis in high-grade serous ovarian carcinoma (HGSOC) [90]. Recent evidence demonstrated that CD73 expression in primary-derived OC promotes stemness and tumor growth and proved that this enzyme acts as an EMT promoter [91], allowing us to recognize CD73 as a promising target for OC.

Eventually, ADO in the extracellular milieu activates ADORA receptors (also known as P1), whose expression has also been characterized in different OCDC lines (e.g., in SKOV-3, CAOV-4, and OVCAR-3). Transcript and protein presence of the four ADORA receptors has been described in the previously mentioned cell lines, with ADORA2B and ADORA3 expression being more abundant; their functionality was demonstrated through cAMP measurement in response to specific agonists [92, 93].

SKOV-3 cell incubation with Apy, ADO, or NECA, an ADORA2 receptor agonist, decreases cell migration. A transcriptional study using a microarray demonstrated that treatment with ADO reduced expression of WNT2, WNT6, WNT10B, and FGF-18, all of which activate signaling pathways involved in EMT in OC [10]. On the other hand, expression of ARPC4 and RAPGEG1 transcripts associated with cytoskeleton rearrangement increased [10]. Regarding ADORA2 expression, it has been proven that the addition of NECA reduces cell viability and promotes apoptosis in CAOV-4 and OVCAR-3 cell lines [92]. Activation of ADORA3 in OCA-derived cell lines is associated with apoptosis induction and G1 phase cell cycle arrest [94]. Altogether, the evidence highlights purinergic signaling as an important regulator in EOC progression.

5. Conclusion

Since Geoffrey Burnstock proposed his purinergic hypothesis in the early 1970s, enormous advances have been achieved in describing the molecular elements that conform the purinergic system and in our understanding of a complex system that is constituted as a continuous metabolic network together with the dynamic events in extracellular signaling. Indeed, the specific actions of purinergic signaling in each system are still being discovered, and its study is a growing field of knowledge. In this chapter, we summarize current knowledge of purinergic signaling in the ovary, where an extensive and specialized expression of purinergic receptors and purine-handling enzymes are observed. The accumulated evidence depicts an emergent and complex system, and at the same time, it raises important questions with deep physiological and pathological implications.

Acknowledgements

This work was funded by PAPIIT-UNAM, number IN201017 to F.G.V.-C. and IN201618 to M.D.-M. and CONACyT number 284-557 to M.D.-M. We thank Jessica González Norris for editing the English version of this manuscript.

Conflict of interest

The authors declare that there is no conflict of interest regarding the publication of this article.

Abbreviations

AC	adenylate cyclase
ADO	adenosine
ADP	adenosine diphosphate
AMP	adenosine monophosphate
AMPK	AMP kinase
Apy	apyrase
ATP	adenosine triphosphate
$[Ca^{2+}]_i$	intracellular concentration of Ca^{2+}
CEO	cumulus-enclosed oocyte
EGFR	epithelial growth factor receptor
egr-1	early growth-1
EMT	epithelial to mesenchymal transition
ENPP	ectonucleotide pyrophosphatase
ENTPD/CD39	ectonucleoside triphosphate diphosphohydrolase
EOC	epithelial ovarian cancer

ERK	extracellular mitogen-regulated kinases
FCS	fetal calf serum
$FADH_2/FAD^+$	reduced/oxidized flavin adenine dinucleotide
$FMNH_2/FMN^+$	reduced/oxidized riboflavin-5′-phosphate
FSH	follicle-stimulating hormone
GLC	granulosa-luteal cells
GPCRs	G protein-coupled receptors
GTP	guanosine triphosphate
hCG	human chorionic gonadotropin
HGSOC	high-grade serous ovarian carcinoma
HPLC	high-performance liquid chromatography
LH	luteinizing hormone
NAADP	nicotinic acid adenine dinucleotide phosphate
$NAD(P)H/NAD(P)^+$	reduced/oxidized nicotinamide adenine dinucleotide phosphate
NECA	5'-N-ethylcarboxamidoadenosine
NK	natural killer
NT5E	ecto-5′-nucleotidase
OC	ovarian cancer
OCDC	ovarian carcinoma-derived cells
OSE	ovarian surface epithelium
PGF2α	prostaglandin F2α
PLD	phospholipase D
PLC	phospholipase C
PKC	protein kinase C
PMSG	pregnant mare serum gonadotropin
8-PTH	8-phenyltheophylline
RyR	ryanodine receptor
SAM	S-adenosylmethionine
SAH	S-adenosylhomocysteine
SERCA	Ca^{2+}-ATPase
TME	tumor microenvironment
UTP	uridine triphosphate

Author details

Mauricio Díaz-Muñoz, Anaí Campos-Contreras, Patricia Juárez-Mercado, Erandi Velázquez-Miranda and Francisco G. Vázquez-Cuevas*
Department of Cellular and Molecular Neurobiology, Neurobiology Institute, National Autonomous University of México, Mexico

*Address all correspondence to: fvazquez@comunidad.unam.mx

References

[1] Freeman ME. Neuroendocrine control of the ovarian cycle of the rat. In: Neill J, editor. Knobil and Neill's Physiology of Reproduction. 3rd ed. Academic Press; 2006. pp. 2328-2388

[2] Monniaux D. Driving folliculogenesis by the oocyte-somatic cell dialog: Lessons from genetic models. Theriogenology. 2016;**86**(1):41-53. DOI: 10.1016/j.theriogenology.2016.04.017

[3] Young JM, McNeilly AS. Theca: The forgotten cell of the ovarian follicle. Reproduction. 2010;**140**:489-504. DOI: 10.1530/REP-10-0094

[4] Cohn M. Structural and chemical properties of ATP and its metal complexes in solution. Annals of the New York Academy of Sciences. 1990; **603**:151-164

[5] Guse AH. Cyclic ADP-ribose. Journal of Molecular and Genetic Medicine. 2000;**78**:26-35

[6] Borea PA, Gessi S, Merighi S, Varani K. Adenosine as a multi-signalling guardian angel in human diseases: When, where and how does it exert its protective effects? Trends in Pharmacological Sciences. 2016;**37**: 419-434. DOI: 10.1016/j.tips.2016.02.006

[7] Camici M, García-Gil M, Tozzi MG. The inside story of adenosine. International Journal of Molecular Sciences. 2018;**19**:E784. DOI: 10.3390/ijms19030784

[8] Oliveira-Giacomelli A, Naaldijk Y, Sardá-Arroyo L, Goncalves MCB, Corréa-Velloso J, Pillat MM, et al. Purinergic receptors in neurological diseases with motor symptoms: Targets for therapy. Frontiers in Pharmacology. 2018;**9**:325. DOI: 10.3389/fphar.2018.00325

[9] Martínez-Ramírez AS, Díaz-Muñoz M, Butanda-Ochoa A, Vázquez-Cuevas FG. Nucleotides and nucleoside signaling in the regulation of the epithelium to mesenchymal transition (EMT). Purinergic Signal. 2017;**13**:1-12. DOI: 10.1007/s11302-016-9550-3

[10] Martínez-Ramírez AS, Díaz-Muñoz M, Battastini AM, Campos-Contreras A, Olvera A, Bergamin L, et al. Cellular migration ability is modulated by extracellular purines in ovarian carcinoma SKOV-3 cells. Journal of Cellular Biochemistry. 2017;**118**: 4468-4478. DOI: 10.1002/jcb.26104

[11] Al-Rashida M, Iqbal J. Therapeutic potential of ecto-nucleoside triphosphate diphosphohydrolase, ectonucleotide pyrophosphatase/phosphodiesterase, ecto-5′-nucleotidase and alkaline phosphatase inhibitors. Medicinal Research Reviews. 2014;**34**: 703-743. DOI: 10.1002/med.21302

[12] Chagoya de Sánchez V, Brunner A, Piña E. In vivo modification of the energy charge in the liver cell. Biochemical and Biophysical Research Communications. 1972;**46**:1441-1445

[13] Hardie DG, Ross FA, Hawley SA. AMP-activated protein kinase: A target for drugs both ancient and modern. Chemistry & Biology. 2012;**19**: 1222-1236. DOI: 10.1016/j.chembiol.2012.08.019

[14] Chagoya de Sánchez V, Hernández-Muñoz R, Suárez J, Vidrio S, Yáñez L, Díaz-Muñoz M. Day-night variations of adenosine and its metabolizing enzymes in the brain cortex of the rat—Possible physiological significance for the energetic homeostasis and the sleep-wake cycle. Brain Research. 1993;**612**:115-121

[15] Velasco-Loyden G, Pérez-Martínez L, Vidrio-Gómez S, Pérez-Carrión JI,

Chagoya de Sánchez V. Cancer chemoprevention by an adenosine derivative in a model of cirrhosis-hepatocellular carcinoma induced by diethylnitrosamine in rats. Tumour Biology. 2017;**39**. DOI: 10.1177/1010428317691190

[16] Yung WS. Metabolism of sulfur-containing amino acids in the liver: A link between hepatic injury and recovery. Biological & Pharmaceutical Bulletin. 2016;**38**:971-974. DOI: 10.1248/bpb.b15-00244

[17] Lazarowski ER. Vesicular and conductive of nucleotide release. Purinergic Signal. 2012;**8**:359-373

[18] Thorn JA, Jarvis SM. Adenosine transporters. General Pharmacology. 1996;**27**:613-620

[19] Falzoni S, Donvito G, Di Virgilio F. Detecting adenosine triphosphate in the pericellular space. Interphase Focus. 2013;**3**. DOI: 10.1098/rsfs.2012.0101

[20] Behrman HR, Polan ML, Ohkawa R, Laufer N, Luborsky JL, Williams AT, et al. Purine modulation of LH action in gonadal cells. Journal of Steroid Biochemistry. 1983;**19**:789-793

[21] Polan ML, DeCherney AH, Haseltine FP, Mezer HC, Behrman HR. Adenosine amplifies follicle-stimulating hormone action in granulosa cells and luteinizing hormone action in luteal cells of rat and human ovaries. The Journal of Clinical Endocrinology and Metabolism. 1983;**56**:288-294

[22] Ohkawa R, Polan ML, Behrman HR. Adenosine differentially amplifies luteinizing hormone-over follicle-stimulating hormone-mediated effects in acute cultures of rat granulosa cells. Endocrinology. 1985;**117**:248-254

[23] Baum MS, Ahrén K. Adenosine potentiates the effects of LH and isoproterenol on luteal cells but not on isolated intact corpora lutea. Acta Physiologica Scandinavica. 1986;**127**: 373-379

[24] Billig H, Rosberg S. Evidence for A2 adenosine receptor-mediated effects on adenylate cyclase activity in rat ovarian membranes. Molecular and Cellular Endocrinology. 1988;**56**:205-210

[25] Billig H, Rosberg S, Johanson C, Ahrén K. Adenosine as substrate and receptor agonist in the ovary. Steroids. 1989;**54**:523-542

[26] Weems CW, Weems YS, Lee CN, Vincent DL. Purine modulation of prostaglandin F2 alpha (PGF2 alpha) induced premature luteolysis in vivo in nonpregnant sheep. Second Messengers and Phosphoproteins. 1992;**14**:1-9

[27] Kamada S, Blackmore PF, Oehninger S, Gordon K, Hodgen GD. Existence of P2-purinoceptors on human and porcine granulosa cells. The Journal of Clinical Endocrinology and Metabolism. 1994;**78**:650-656. DOI: 10.1210/jcem.78.3.8126137

[28] Morley P, Vanderhyden BC, Tremblay R, Mealing GA, Durkin JP, Whitfield JF. Purinergic receptor-mediated intracellular Ca^{2+} oscillations in chicken granulosa cells. Endocrinology. 1994;**134**:1269-1276. DOI: 10.1210/endo.134.3.8119167

[29] Squires PE, Lee PS, Yuen BH, Leung PC, Buchan AM. Mechanisms involved in ATP-evoked Ca^{2+} oscillations in isolated human granulosa-luteal cells. Cell Calcium. 1997;**21**:365-374

[30] Tai CJ, Kang SK, Cheng KW, Choi KC, Nathwani PS, Leung PC. Expression and regulation of P2U-purinergic receptor in human granulosa-luteal cells. The Journal of Clinical Endocrinology and Metabolism. 2000; **85**:1591-1597. DOI: 10.1210/jcem.85.4.6558

[31] Tai CJ, Kang SK, Choi KC, Tzeng CR, Leung PC. Antigonadotropic action ofadenosine triphosphate in human granulosa-luteal cells: involvement of protein kinase C alpha. The Journal of Clinical Endocrinology and Metabolism. 2001;**86**:3237-3242. DOI: 10.1210/jcem.86.7.7691

[32] Morales-Tlalpan V, Arellano RO, Díaz-Muñoz M. Interplay between ryanodine and IP3 receptors in ATP-stimulated mouse luteinized-granulosa cells. Cell Calcium. 2005;**37**:203-213. DOI: 10.1016/j.ceca.2004.10.001

[33] Díaz-Muñoz M, de la Rosa Santander P, Juárez-Espinosa AB, Arellano RO, Morales-Tlalpan V. Granulosa cells express three inositol 1,4,5-trisphosphate receptor isoforms: cytoplasmic and nuclear Ca^{2+} mobilization. Reproductive Biology and Endocrinology. 2008;**6**:60. DOI: 10.1186/1477-7827-6-60

[34] Bintig W, Baumgart J, Walter WJ, Heisterkamp A, Lubatschowski H, Ngezahayo A. Purinergic signalling in rat GFSHR-17 granulosa cells: An in vitro model of granulosa cells in maturing follicles. Journal of Bioenergetics and Biomembranes. 2009; **41**:85-94. DOI: 10.1007/s10863-009-9199-5

[35] Tai CJ, Kang SK, Leung PC. Adenosine triphosphate-evoked cytosolic calcium oscillations in human granulosa-luteal cells: Role of protein kinase C. The Journal of Clinical Endocrinology and Metabolism. 2001; **86**:773-777. DOI: 10.1210/jcem.86.2.7231

[36] Tai CJ, Kang SK, Tzeng CR, Leung PC. Adenosine triphosphate activates mitogen-activated protein kinase in human granulosa-luteal cells. Endocrinology. 2001;**142**:1554-1560. DOI: 10.1210/endo.142.4.8081

[37] Tai CJ, Chang SJ, Leung PC, Tzeng CR. Adenosine 5′-triphosphate activates nuclear translocation of mitogen-activated protein kinases leading to the induction of early growth response 1 and raf expression in human granulosa-luteal cells. The Journal of Clinical Endocrinology and Metabolism. 2004; **89**:5189-5195. DOI: 10.1210/jc.2003-032111

[38] Bao R, Xu P, Wang Y, Li J, Zhang J, Wang J, et al. P2Y6 purinergic receptor regulates steroid synthesis and proliferation of ovarian luteal cells. International Journal of Clinical and Experimental Medicine. 2017;**10**(8): 11508-11518

[39] Park DW, Cho T, Kim MR, Kim YA, Min CK, Hwang KJ. ATP-induced apoptosis of human granulosa luteal cells cultured in vitro. Fertility and Sterility. 2003;**80**:993-1002

[40] Tai CJ, Chang SJ, Chien LY, Leung PC, Tzeng CR. Adenosine triphosphate induces activation of caspase-3 in apoptosis of human granulosa-luteal cells. Endocrine Journal. 2005;**52**: 327-335

[41] Coutinho-Silva R, Persechini PM, Bisaggio RD, Perfettini JL, Neto AC, Kanellopoulos JM, et al. P2Z/P2X7 receptor-dependent apoptosis of dendritic cells. The American Journal of Physiology. 1999;**276**:C1139-C1147

[42] Wang J, Liu S, Nie Y, Wu B, Wu Q, Song M, et al. Activation of P2X7 receptors decreases the proliferation of murine luteal cells. Reproduction, Fertility, and Development. 2015;**27**: 1262-1271. DOI: 10.1071/RD14381

[43] Richards JS, Ren YA, Candelaria N, Adams JE, Rajkovic A. Ovarian follicular theca cell recruitment, differentiation and impact on fertility: 2017 update. Endrocrine Reviews. 2018;**39**:1-20. DOI: 10.1210/er.2017-00164

[44] Chu YL, Xu YR, Yang WX, Sun Y. The role of FSH and TGF-β superfamily

in follicle atresia. Aging. 2018;**10**: 305-321. DOI: 10.18632/aging.101391

[45] Lansdown A, Rees DA. The sympathetic nervous system in polycystic ovary syndrome: A novel therapeutic target? Clinical Endocrinology. 2012;**77**:791-801. DOI: 10.1111/cen.12003

[46] Magoffin DA, Weitsman SR. Insulin-like growth factor-I regulation of luteinizing hormone (LH) receptor messenger ribonucleic acid expression and LH-stimulated signal transduction in rat ovarian theca interstitial cells. Biology of Reproduction. 1994;**51**: 766-775. DOI: 10.1095/biolreprod51.4.766

[47] Salvetti NR, Panzani CG, Gimeno EJ, Neme LG, Alfaro NS, Ortega HH. An imbalance between apoptosis and proliferation contributes to follicular persistence in polycystic ovaries in rats. Reproductive Biology and Endocrinology. 2009;**7**:68. DOI: 10.1186/1477-7827-7-68

[48] Magoffin DA. Ovarian theca cell. The International Journal of Biochemistry & Cell Biology. 2005;**37**: 1344-1349. DOI: 10.1016/j.biocel.2005.01.016

[49] Uchida S. Sympathetic regulation of estradiol secretion from the ovary. Autonomic Neuroscience. 2015;**187**: 27-35. DOI: 10.1016/j.autneu.2014.10.023

[50] Burnstock G. Purinergic signalling: Therapeutic developments. Frontiers in Pharmacology. 2017;**8**:661. DOI: 10.3389/fphar.2017.00661

[51] Aleu J, Martín-Satué M, Navarro P, Pérez de Lara I, Bahima L, Marsal J, et al. Release of ATP induced by hypertonic solutions in Xenopus oocytes. The Journal of Physiology. 2003;**547**:209-219

[52] Burnstock G. Purinergic signalling in the reproductive system in health and disease. Purinergic Signal. 2014;**10**: 157-187. DOI: 10.1007/s11302-013-9399-7

[53] Vázquez-Cuevas FG, Juárez B, Garay E, Arellano RO. ATP-induced apoptotic cell death in porcine ovarian theca cells through P2X7 receptor activation. Molecular Reproduction and Development. 2006;**73**:745-755

[54] Vázquez-Cuevas FG, Zárate-Díaz EP, Garay E, Arellano RO. Functional expression and intracellular signaling of UTP-sensitive P2Y receptor in theca-interstitial cells. Reproductive Biology and Endocrinology. 2010;**8**:88. DOI: 10.1186/1477-7827-8-88

[55] Arellano RO, Garay E, Vázquez-Cuevas F. Functional interaction between native G protein-coupled purinergic receptors in Xenopus follicles. Proceedings of the National Academy of Sciences of the United States of America. 2009;**106**: 16680-16685. DOI: 10.1073/pnas.0905811106

[56] Eppig J, Ward-Bailey D, Coleman L. Hypoxanthine concentrations and adenosine in murine ovarian follicular and activity in maintaining oocyte meiotic fluid. Biology of Reproduction. 1985;**33**:1041-1049. DOI: 10.1095/biolreprod33.5.1041

[57] Downs SM, Coleman DL, Eppig JJ. Maintenance of murine oocyte meiotic arrest: Uptake and metabolism of hypoxanthine and adenosine by cumulus cell-enclosed and denuded oocytes. Developmental Biology. 1986; **117**:174-183. DOI: 10.1016/0012-1606(86)90359-3

[58] Arellano RO, Martínez-Torres A, Garay E. Ionic currents activated via purinergic receptors in the cumulus cell-enclosed mouse oocyte. Biology of Reproduction. 2002;**67**(3):837-846. DOI: 10.1095/biolreprod.102.003889

[59] Webb RJ, Bains H, Cruttwell C, Carroll J. Gap-junctional communication in mouse cumulus-oocyte complexes: Implications for the mechanism of meiotic maturation. Reproduction. 2002;**123**:41-52 DOI: rep.0.1230041

[60] Eppig JJ. The relationship between parthenogenetic embryonic development and cumulus cell-oocyte intercellular coupling during oocyte meiotic maturation. Gamete Research. 1982;**5**:229-237. DOI: 10.1002/mrd.1120050303

[61] Downs SM, Coleman DL, Ward-bailey PF, Eppig JJ. Hypoxanthine is the principal inhibitor of murine oocyte maturation in a low molecular weight fraction of porcine follicular fluid Developmental Biology. Proceedings of the National Academy of Sciences. 1985;**82**(2):454-458. DOI: 10.1073/pnas.82.2.454

[62] Murdoch WJ, McDonnel AC. Roles of the ovarian surface epithelium in ovulation and carcinogenesis. Reproduction. 2002;**123**(6):743-750. DOI: 10.1530/reprod/123.6.743

[63] Szotek PP, Chang HL, Brennand K, et al. Normal ovarian surface epithelial label-retaining cells exhibit stem/progenitor cell characteristics. Proceedings of the National Academy of Sciences. 2008;**105**(34):12469-12473. DOI: 10.1073/pnas.0805012105

[64] Auersperg N, Wong AST, Choi K, Kang SK, Leung PCK. Ovarian surface epithelium: Biology, endocrinology and pathology. Endocrine Reviews. 2001;**22**(2):255-288. DOI: 10.1210/er.22.2.255

[65] Ng A, Barker N. Ovary and fimbrial stem cells: Biology, niche and cancer origins. Nature Reviews. Molecular Cell Biology. 2015;**16**(10):625-638. DOI: 10.1038/nrm4056

[66] Atwood CS, Vadakkadath Meethal S. The spatiotemporal hormonal orchestration of human folliculogenesis, early embryogenesis and blastocyst implantation. Molecular and Cellular Endocrinology. 2016;**430**:33-48. DOI: 10.1016/j.mce.2016.03.039

[67] Choi K-C. Follicle-stimulating hormone activates mitogen-activated protein kinase in preneoplastic and neoplastic ovarian surface epithelial cells. The Journal of Clinical Endocrinology and Metabolism. 2002; **87**(5):2245-2253. DOI: 10.1210/jc.87.5.2245

[68] Murdoch J, Van Kirk EA, Murdoch WJ. Hormonal control of urokinase plasminogen activator secretion by sheep ovarian surface epithelial cells. Biology of Reproduction. 1999;**61**(6): 1487-1491

[69] Wright JW, Pejovic T, Lawson M, Jurevic L, Hobbs T, Stouffer RL. Ovulation in the absence of the ovarian surface epithelium in the primate. Biology of Reproduction. 2010;**82**(3): 599-605. DOI: 10.1095/biolreprod.109.081570

[70] Vázquez-Cuevas FG, Cruz-Rico A, Garay E, Pérez-Montiel D, Juárez B, Arellano RO. Differential expression of the P2X7 receptor in ovarian surface epithelium during the oestrous cycle in the mouse. Reproduction, Fertility, and Development. 2013:971-984. DOI: 10.1071/RD12196

[71] Lengyel E. Ovarian cancer development and metastasis. The American Journal of Pathology. 2010; **177**(3):1053-1064. DOI: 10.2353/ajpath.2010.100105

[72] Globocan 2012 [Internet]. 2012. Available from: http://globocan.iarc.fr/Default.aspx. [Accessed: 2018-03-24]

[73] Karst AM, Drapkin R. Ovarian cancer pathogenesis: A model in

evolution. Journal of Oncology. 2010; **2010**:932371. DOI: 10.1155/2010/932371

[74] Matulonis UA, Sood AK, Fallowfield L, Howitt BE, Sehouli J, Karlan BY. Ovarian cancer. Nature Review Disease Primers. 2016;**2**:16061. DOI: 10.1038/nrdp.2016.61

[75] Ahmed N, Stenvers KL. Getting to know ovarian cancer ascites: Opportunities for targeted therapy-based translational research. Frontiers in Oncology. 2013;**3**:256. DOI: 10.3389/fonc.2013.00256

[76] Klymenko Y, Kim O, Stack MS. Complex determinants of epithelial: Mesenchymal phenotypic plasticity in ovarian cancer. Cancers (Basel). 2017;**9**. DOI: 10.3390/cancers9080104

[77] Di Virgilio F, Adinolfi E. Extracellular purines, purinergic receptors and tumor growth. Oncogene. 2017;**36**:293-303. DOI: 10.1038/onc.2016.206

[78] Nakajima EC, Van Houten B. Metabolic symbiosis in cancer: Refocusing the Warburg lens. Molecular Carcinogenesis. 2013;**52**:329-337. DOI: 10.1002/mc.21863

[79] Ohta A, Gorelik E, Prasad SJ, Ronchese F, Lukashev D, Wong MK, et al. A2A adenosine receptor protects tumors from antitumor T cells. Proceedings of the National Academy of Sciences of the United States of America. 2006;**103**:13132-13137. DOI: 10.1073/pnas.0605251103

[80] Pellegatti P, Raffaghello L, Bianchi G, Piccardi F, Pistoia V, Di Virgilio F. Increased level of extracellular ATP at tumor sites: in vivo imaging with plasma membrane luciferase. PLoS One. 2008;**3**: e2599. DOI: 10.1371/journal.pone.0002599

[81] Jiang T, Xu X, Qiao M, Li X, Zhao C, Zhou F, et al. Comprehensive evaluation of NT5E/CD73 expression and its prognostic significance in distinct types of cancers. BMC Cancer. 2018;**18**:267. DOI: 10.1186/s12885-018-4073-7

[82] Batra S, Fadeel I. Release of intracellular calcium and stimulation of cell growth by ATP and histamine in human ovarian cancer cells (SKOV-3). Cancer Letters. 1994;**77**:57-63. DOI: 10.1016/0304-3835(94)90348-4

[83] Popper LD, Batra S. Calcium mobilization and cell proliferation activated by extracellular ATP in human ovarian tumour cells. Cell Calcium. 1993;**14**:209-218. DOI: 10.1016/0143-4160(93)90068-H

[84] Schultze-Mosgau A, Katzur AC, Arora KK, Stojilkovic SS, Diedrich K, Ortmann O. Characterization of calcium-mobilizing, purinergic P2Y(2) receptors in human ovarian cancer cells. Molecular Human Reproduction. 2000; **6**:435-442

[85] Choi K-C, Tai C-J, Tzeng C-R, Auersperg N, Leung PCK. Adenosine triphosphate activates mitogen-activated protein kinase in pre-neoplastic and neoplastic ovarian surface epithelial cells. Biology of Reproduction. 2003;**68**:309-315

[86] Adinolfi E, Raffaghello L, Giuliani AL, Cavazzini L, Capece M, Chiozzi P, et al. Expression of P2X7 receptor increases in vivo tumor growth. Cancer Research. 2012;**72**:2957-2969. DOI: 10.1158/0008-5472.CAN-11-1947

[87] Vázquez-Cuevas FG, Martínez-Ramírez AS, Robles-Martínez L, Garay E, García-Carrancá A, Pérez-Montiel D, et al. Paracrine stimulation of P2X7 receptor by ATP activates a proliferative pathway in ovarian carcinoma cells. Journal of Cellular Biochemistry. 2014; **115**:1955-1966. DOI: 10.1002/jcb.24867

[88] Martínez-Ramírez AS, Garay E, García-Carrancá A, Vázquez-Cuevas

FG. The P2RY2 receptor induces carcinoma cell migration and EMT through cross-talk with epidermal growth factor receptor. Journal of Cellular Biochemistry. 2016;**117**: 1016-1026. DOI: 10.1002/jcb.25390

[89] Häusler SF, del Barrio IM, Diessner J, Stein RG, Strohschein J, Hönig A, et al. Anti-CD39 and anti-CD73 antibodies A1 and 7G2 improve targeted therapy in ovarian cancer by blocking adenosine-dependent immune evasion. American Journal of Translational Research. 2014; **6**:129-139

[90] Turcotte M, Spring K, Pommey S, Chouinard G, Cousineau I, George J, et al. CD73 is associated with poor prognosis in high-grade serous ovarian cancer. Cancer Research. 2015;**75**: 4494-4503. DOI: 10.1158/0008-5472.CAN-14-3569

[91] Lupia M, Angiolini F, Bertalot G, Freddi S, Sachsenmeier KF, Chisci E, et al. CD73 regulates stemness and epithelial-mesenchymal transition in ovarian cancer-initiating cells. Stem Cell Reports. 2018;**10**:1412-1425. DOI: 10.1016/j.stemcr.2018.02.009

[92] Sima H, Panjehpour M, Aghaei M, Shabani M. Activation of A2b adenosine receptor regulates ovarian cancer cell growth: involvement of Bax/Bcl-2 and caspase-3. Biochemistry and Cell Biology. 2015;**93**:321-329. DOI: 10.1139/bcb-2014-0117

[93] Hajiahmadi S, Panjehpour M, Aghaei M, Mousavi S. Molecular expression of adenosine receptors in OVCAR-3, Caov-4 and SKOV-3 human ovarian cancer cell lines. Research in Pharmaceutical Sciences. 2015;**10**:43-51

[94] Joshaghani HR, Jafari SM, Aghaei M, Panjehpour M, Abedi H. A3 adenosine receptor agonist induce G1 cell cycle arrest via cyclin D and cyclin-dependent kinase 4 pathways in OVCAR-3 and Caov-4 cell lines. Journal of Cancer Research and Therapeutics. 2017;**13**:107-112. DOI: 10.4103/0973-1482.199381

Chapter 4

Phosphorylation of NF-κB in Cancer

Matthew Martin, Antja-Voy Hartley, Jiamin Jin, Mengyao Sun and Tao Lu

Abstract

The proinflammatory transcription factor nuclear factor-κB (NF-κB) has emerged as a central player in inflammatory responses and tumor development since its discovery three decades ago. In general, aberrant NF-κB activity plays a critical role in tumorigenesis and acquired resistance to chemotherapy. This aberrant NF-κB activity frequently involves several post-translational modifications of NF-κB, including phosphorylation. In this chapter, we will specifically cover the phosphorylation sites reported on the p65 subunit of NF-κB and their relationship to cancer. Importantly, phosphorylation is catalyzed by different kinases using adenosine triphosphate (ATP) as the phosphorus donor. These kinases are frequently hyperactive in cancers and thus may serve as potential therapeutic targets to treat different cancers.

Keywords: kinase, NF-κB, phosphorylation, post-translational modifications

1. Introduction

1.1 Basic nuclear factor-κB (NF-κB) family and signaling pathways

So what is NF-κB? In mammals, NF-κB is a collective term for a small family of dimeric transcription factors [comprising p65 (RelA) and RelB, c-Rel, p50/p105 (NF-κB1), and p52/p100 (NF-κB2)]. All NF-κB proteins share a Rel homology domain (RHD), which is responsible for DNA binding and dimerization. Only p65, RelB, and c-Rel contain potent transactivation domains within sequences from C-terminal to the RHD. Therefore, p50 and p52 cannot act as transcriptional activators by themselves. Dimers of these two proteins have been reported to repress NF-κB-dependent transcription *in vivo*, most likely by competing with other transcriptionally active dimers. These proteins form homo- and heterodimers, and their activity is regulated by the canonical or alternative pathways described as following [1]. A simple diagram of canonical NF-κB signaling, which will be the focus of this chapter, is shown in **Figure 1**. The canonical pathway is activated by multiple stimuli, including proinflammatory cytokines (e.g., tumor necrosis factor, TNF; interleukin 1, IL-1), and the components of the bacterial wall (Lipopolysaccharide, LPS). Exterior signals lead to the phosphorylation and degradation of the inhibitory complex IκB, which is modulated by the IκB kinase (IKK), and its degradation allows for the release of the typical NF-κB heterodimer, p65/p50, to translocate into the nucleus. NF-κB binds to its cognate DNA elements and can transcriptionally activate different target genes among which 200–500 genes have been implicated in cell survival/apoptosis, cell growth, immune response, and inflammation [2].

The alternative or noncanonical pathway is activated by the members of the TNF cytokine family, such as B-cell activating factor (BAFF), cluster of differentiation 40 ligand (CD40L), receptor activator of nuclear factor-κB ligand (RANKL), and lymphotoxin-β2 (LTβ2), and requires recruitment of the p52/RelB dimers to activate transcription. Firstly, activation of NIK (NF-κB-inducing kinase) leads to

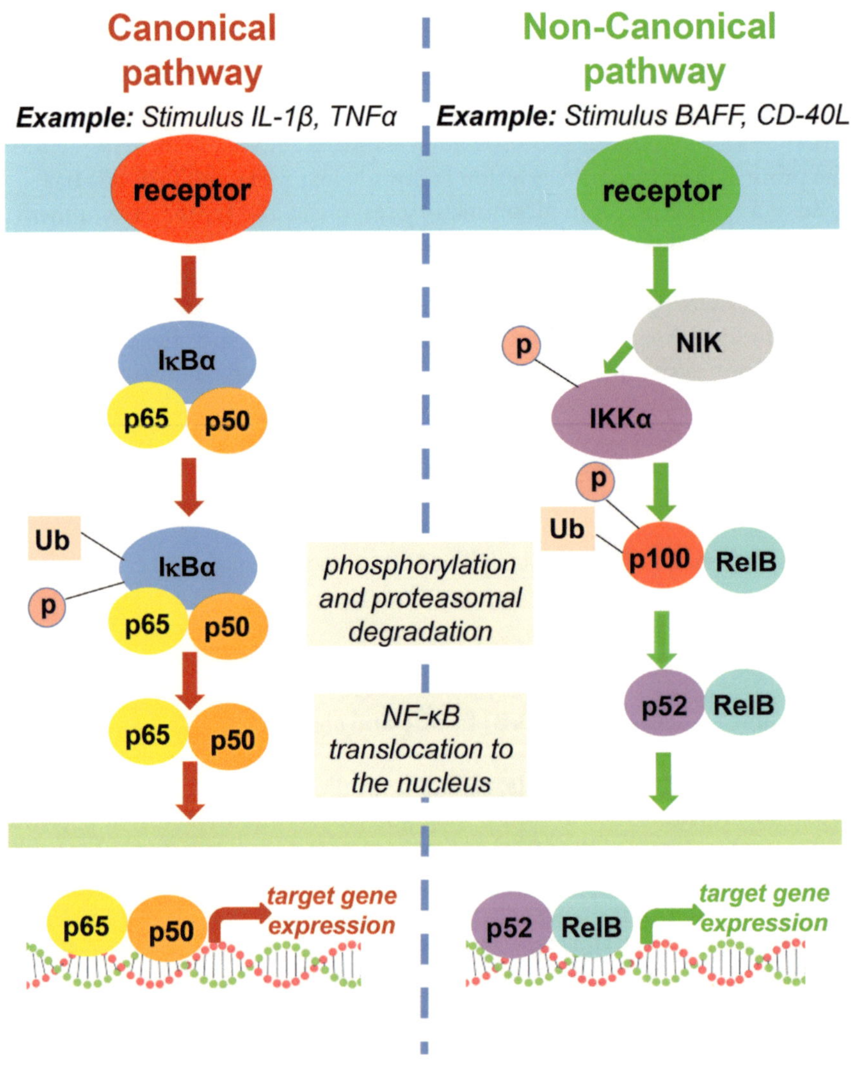

Figure 1.
Pathway of canonical and non-canonical NF-κB signaling. Under the canonical pathway of NF-κB signaling, activation of the NF-κB is initiated by a stimulus resulting in phosphorylation and subsequent proteasomal degradation of IκBα. This allows the release of the p65/p50 heterodimer into the nucleus, where they can bind to their cognate DNA elements and promote NF-κB target gene expression [1]. On the other hand, under noncanonical activation of NF-κB, NIK (NF-κB-inducing kinase) leads to activation of IKKα in this pathway. Subsequent phosphorylation of the NF-κB1 precursor molecule, p100, triggers partial proteolysis giving rise to p52, which preferentially dimerizes with RelB. This allows translocation of p52/RelB to the nucleus where they can bind to cognate DNA elements and promote gene transcription [1]. Figure adapted and simplified from Hoesel et al. [103].

activation of IKKα in this pathway. This event leads to the subsequent phosphorylation of the NF-κB1 precursor molecule, p100, and triggers partial proteolysis to give rise to p52, which preferentially dimerizes with RelB [1]. The p52/RelB heterodimer then translocates to the nucleus where they can bind to cognate DNA elements and promote gene transcription (**Figure 1**).

1.2 Important role of NF-κB in cancer

NF-κB was first discovered by Dr. Ranjan Sen in 1986 [3]. This family of transcription factors plays important roles in the regulation of apoptosis, proliferation, inflammation, and immune response in both normal and cancer cells. Generally, in normal cells, the central transcription factor NF-κB is transiently activated in response to certain stimuli. However, cancer cells usually exhibit sustained activation of NF-κB [4, 5] which significantly contributes to their survival. Moreover, NF-κB activity plays critical roles in many of the well-known "hallmarks" of cancer, via its regulation of target genes involved in tumor cell proliferation, suppression of apoptosis, activation of angiogenesis as well as induction of the epithelial-to-mesenchymal transition (EMT) phenotype, a critical step in metastasis [6]. Constitutively active NF-κB has been found in many types of cancer. For instance, in thyroid cancer, oncogenic proteins including "rearranged during transfection" (RET), "Rat Sarcoma" (RAS), and "v-Raf murine sarcoma viral oncogene homolog B" (BRAF) were shown to induce NF-κB activation, which in turn activated proliferative and antiapoptotic signaling pathways [7]. Moreover, in renal cell carcinoma (RCC), NF-κB is constitutively activated. The phosphorylated p65, a major subunit of NF-κB, exhibited a significant increase in the RCC samples compared with corresponding normal tissues [8]. Furthermore, Nogueira et al. showed that in glioblastoma (GBM), deletion of IκB showed a phenotype similar to that of epidermal growth factor receptor (EGFR) amplification in the pathogenesis of GBM. This was also correlated with low survival rates in affected patients [2].

Importantly, our laboratory also found that in colon cancer, NF-κB can be activated by the Y-box protein 1 (YBX1), a critical event correlated with increased colon cancer cell proliferation and anchorage-independent growth [9, 10]. Additionally, in pancreatic cancer, a mutant oncogenic KrasG12D (glycine to aspartic acid) mutation induced positive feedback loops of interleukin 1α (IL-1α) and p62 expression to sustain constitutive IKKβ (inhibitor of NF-κB kinase subunit β)/NF-κB activation [11]. In breast cancer, moderately elevated NF-κB led to chronic inflammatory conditions that result in some cells escaping immune surveillance [12]. Additionally, NF-κB activation was shown to upregulate the expression of cyclin D1, cyclin-dependent kinase 2 (CDK2), and c-Myc, which drives cell cycle progression and causes uncontrolled cell proliferation [13]. Moreover, in breast cancer, NF-κB has been shown to induce and stabilize the expression of EMT markers (Snail and twist-related protein1) [14], a pivotal process in tumor metastasis. In addition to the role of NF-κB in solid tumors as described thus far, Gasparini et al. has thoroughly reviewed the important tumorgenic role of NF-κB in hematological malignancies, including acute lymphocytic lymphoma (ALL), acute myeloid leukemia (AML), chronic lymphocytic leukemia (CLL), chromic myeloid leukemia (CML), B lymphomas, diffuse large B-cell lymphomas (DLBCLs), Hodgkin's lymphoma, adult T-cell lymphomas (ATLL), anaplastic large-cell lymphomas (ALCL), and multiple myeloma, which will not be further discussed in this chapter [15]. Overall, these studies highlight a prominent role of dysregulated NF-κB in multiple aspects of cancer progression in both solid tumors and hematological malignancies.

2. Role of phosphorylation of NF-κB in cancer

2.1 Phosphorylation of the p65 subunit of NF-κB and the role of its kinases in cancer

Phosphorylation is a critical modification for NF-κB activation and plays an indispensable role in the regulation of its target genes. Moreover, as mentioned above, many of these target genes contribute to the hallmarks of cancer such as cellular proliferation, antiapoptosis as well as enhanced angiogenesis via vascular endothelial growth factor expression, among others [6]. Thus, understanding how phosphorylation of NF-κB contributes to these cancer phenotypes is a critical step in effectively limiting NF-κB activity [16]. Generally speaking, phosphorylation requires phosphorus which is supplied by the donor molecule adenosine tri-phosphate (ATP). Although several members of the NF-κB family of proteins are reported to be subjected to phosphorylation, p65 stands out as the most frequently modified subunit (**Table 1**). Furthermore, scientists have found that p65 can be phosphorylated by a variety of different kinases, some of which are themselves frequently overactive in cancer. For instance, p65 phosphorylation at serine 536 (S536) by IKKβ has been shown to be critical for TNFα-induced transformation of mouse epidermal cells [17]. Additional studies have also reported a role for p65 S536 phosphorylation in mediating expression of matrix metalloproteinase 1 (MMP-1) in lymphomas, wherein high MMP-1 expression correlated with lymphatic invasion and lymph node metastasis [18]. Another study with an immortalized prostate cell line, PNT1a, showed a role for phosphorylated S536 of p65 in cell motility and transformation [19].

As mentioned above, phosphorylation events on NF-κB are mediated by a variety of kinases. It is therefore unsurprising that the action of these kinases has been tightly regulated to maintain normal cellular function. However, deregulation

Known phosphorylation sites of p65	Cancers involved	Cell line discovered	References
S205	No cancers currently known	HEK 293 cells	[104]
T254	Breast cancer	BT20 and MCF-7 cells	[105]
S529	Breast cancer	HeLa cells	[33]
S536	Bone cancer	HeLa and BC-3 cells	[35]
S276	Head and neck cancers, breast cancer	HNSCC cells	[34]
S281	No cancers currently known	MEF cells	[106]
S311	No cancers currently known	HEK 293 and MEF cells	[107]
T435	No cancers currently known	SiHa cells	[108]
S468	No cancers currently known	HeLa cells	[109]
T505	No cancers currently known	NARF2 and Hs68 cells	[110]
S535	No cancers currently known	HeLa cells	[111]
S316	No cancers currently known	HEK 293 cells	[41]
S547	No cancers currently known	HEK 293 cells	[112]

Table 1.
List of known phosphorylation sites on the p65 subunit of NF-κB and their relationship to cancer.

of kinase activity can have detrimental downstream effects, which also involves the aberrant activation of NF-κB and its target genes to promote a cancer phenotype. Several kinases have been shown to have critical roles in the regulation of the p65 subunit of NF-κB. For example, glycogen synthase kinase 3 beta (GSK3β) and TRAF-associated NF-κB activator TBK1 ((TANK)-binding kinase 1) have been shown to be critical activators of NF-κB signaling [20, 21] by targeting p65 for phosphorylation on S536. This phosphorylation leads to enhanced NF-κB transactivation both *in vitro* and *in vivo* [22–24]. Another well-known kinase involved in modifying p65 is protein kinase A (PKAc). PKAc is typically activated following IκBα-degradation, leading to PKAc-mediated phosphorylation of p65 on S276 [25]. This phosphorylation event causes recruitment of histone acetyltransferases including cAMP response element-binding (CREB)-binding protein (CBP) and p300. The net effect is displacement of p50-histone deacetylase (HDAC)-1 complex from DNA, which increases p65 transactivation ability [26, 27]. Other kinases can also phosphorylate p65 at S276. These include mitogen- and stress-activated protein kinase-1 and 2 (MSK1, 2), proto-oncogene serine/threonine-protein kinase PIM-1 (PIM-1), ribosomal s6 kinase (RSK) p90, and protein kinase C α (PKCα) [28–31]. Moreover, casein kinase II (CK2), which phosphorylates p65 on S529, has been implicated in breast cancer [32, 33]. Another study demonstrated a role for p65 S276 phosphorylation by protein kinase A (PKA) in promoting a malignant phenotype in head and neck squamous cell carcinoma (HNSCC). Here, the authors found that S276 phosphorylation was prevalent in the nucleus of HNSCC samples but cytoplasmic in normal mucosa. Furthermore, this TNF-α-induced nuclear p65-S276 phosphorylation was significantly inhibited by the PKA inhibitor H-89, which in turn suppressed NF-κB activity, target gene expression, cell proliferation, and induced cell death via G1/S phase arrest [34]. Other p65-targeting kinases implicated in cancer include cell division protein kinase 6 (CDK6) and PIM-1, which phosphorylate p65 at S536 and S276, respectively [29, 35]. Both PIM1 and CDK6 have been shown to be overexpressed in a variety of cancers including hematological cancers, prostate cancer, pancreatic cancer, gastric cancer, head and neck cancer, liver cancer, glioblastomas, medulloblastomas, colon cancer, and lung cancers [36–39]. However, their exact roles in regulating p65 phosphorylation in these cancers are yet to be understood. In gastric cancer cells, Aurora Kinase A (AURKA) was also shown to phosphorylate S536 in *in vivo* and *in vitro* models [40] whereby overexpression of AURKA induced a significant increase in NF-κB p65 and phospho-p65 (S536) protein levels. Interestingly, protein kinase C δ (PKCδ), a member of novel PKC isoforms, has also been implicated in a number of cancers including breast, pancreatic, prostate, and melanoma tumor cells but has been shown to regulate p65 transactivation in a phosphorylation-independent manner [41]. Our laboratory has also recently reported a novel phosphorylation site on S316 of p65, a modification mediated by the kinase CKII [41]. We showed that S316 phosphorylation was necessary for NF-κB activation and target gene expression. Collectively, these examples indicate the importance and sophistication of p65 phosphorylation and their corresponding kinases in regulating NF-κB signaling in the context of cancer.

2.2 Importance and effects of phosphorylation of p65 in modulating chemoresistance

Several studies have indicated a role for NF-κB hyperactivity in the development of resistance to chemotherapeutics via downregulation of antisurvival and upregulation of prosurvival target genes and pathways [42–44]. In one study for example, gemcitabine-resistant pancreatic cancer cells were rendered sensitive to gemcitabine upon knockdown of p65 [42]. These and other accounts of NF-κB-mediated chemoresistance have been extensively reviewed by others such as Li, Sethi, and Godwin et al., which will not be further discussed in this chapter [45, 46]. However, the specific contribution of dysregulated p65 phosphorylation to chemoresistance is less well understood and requires further exploration. Nonetheless, a few reports suggest that upstream kinases involved in chemoresistance can modulate p65 phosphorylation levels in this context. For instance, siRNA-mediated depletion of IKKα in HT1080 human fibrosarcoma cells was shown to decrease phosphorylation of p65 in response to doxorubicin, thus severely impairing the ability of doxorubicin to initiate NF-κB DNA-binding activity. These findings suggest that IKKα plays a critical role in NF-κB-mediated chemoresistance in response to doxorubicin and potentially serves as a therapeutic target for improving chemotherapeutic response [47]. Other studies have shown that p65, in a hyperphosphorylated state, can be correlated with resistance to thymidylate synthases and irinotecan in stomach and colon cancers, respectively [44, 48, 49]. Doxorubicin resistance in lung cancer has also been correlated with p65 S536 phosphorylation states [47]. Additionally, multiple myelomas have exhibited increased p65 S536 phosphorylation within melphalan- or doxorubicin-resistant cells [50].

3. p65 modifying kinases as potential therapeutic targets

3.1 Current therapeutics used to treat cancers with constitutive NF-κB activity

The NF-κB pathway is widely considered an attractive therapeutic target in a broad range of cancers. Yet, despite the efforts to develop NF-κB inhibitors, none has been clinically approved. This is largely due to immune-related toxicities associated with global NF-κB suppression [51]. Furthermore, the high complexity of the NF-κB signaling network presents another unique challenge for developing specific NF-κB inhibitors. To further complicate matters, some standard anticancer agents can inadvertently activate the NF-κB pathway via induction of proinflammatory cytokines such as IL-1β and TNF-α and cellular stressors such as reactive oxygen species (ROS), or by activating DNA-repair mechanisms [52]. Finally, constitutive NF-κB activity can also be achieved via secreted cytokines and chemokines from inflammatory cells within the tumor microenvironment [52]. Taken together, consideration of all these factors is imperative when strategizing the development of the most effective and least toxic anticancer agents.

Currently, the use of NF-κB inhibitors has mainly been combined with other agents [47, 50]. Some of these combinatorial therapies have shown promising effectiveness and have been made it as far as the clinical trial phase. For example, combination of irinotecan with the proteasome and NF-κB inhibitor bortezomib was shown to increase sensitivity of colon cancer cells to irinotecan [53]. A separate study showed that bortezomib could sensitize non-small cell lung cancer (NSCLC) cells to sodium butyrate, which acts to inhibit histone deacetylases [54]. Moreover, several clinical trials testing the efficacy of inhibitors against IKK to target solid tumors have been undertaken. For example, perturbation of IKKβ with the inhibitory ML120B led to synergistic enhancement of vincristine cytotoxicity in lymphoma. These results implicate IKK disruption using inhibitors as a useful adjunct therapy with standard chemotherapeutics. Other attempted trials using IKK inhibitors, such as CHS-828, EB-1627, and IMD-1041, as single or combinatorial agents unfortunately produced toxicity concerns for patients [55–57].

Other examples of combinatorial therapies include the use of NF-κB inhibitors Bay11-7082 and sulfasalazine in combination with more commonly used chemotherapeutics such as 5-fluorouracil and cisplatin to synergistically reduce colon cancer cell growth [58]. Other indirect means of targeting NF-κB such as inhibition of upstream kinases have also shown promise. For instance, one study using pancreatic cancer cells showed that inhibition of GSK3β by a small molecule inhibitor reduced phosphorylation of p65 at S536 resulting in decreased NF-κB activity and cell growth [59]. Another study with the chemical compound ursolic acid showed reduced p65 phosphorylation via inhibition of IKKβ, which impaired overall cell growth in leukemia cell lines [60]. Other studies with proteasome inhibitors, including Tosyl phenylalanyl chloromethyl ketone (TPCK) and Tosyl-L-lysyl-chloromethane hydrochloride (TLCK), demonstrated that not only do these inhibitors target IKKβ, but they were also able to reduce overall phosphorylation levels of NF-κB [61, 62]. In summary, these studies suggest there may be many benefits to targeting hyperactive NF-κB signaling and, in particular, the kinases that regulate NF-κB in various cancers.

3.2 Benefits and pitfalls for targeting kinases in cancers

The development of small-molecule kinase inhibitors for the treatment of cancer has continued to be of intense interest. Notably, many inhibitors have received FDA approval with approximately another 150 are in preclinical and clinical phase trials. Despite these important advances, many factors have confounded the clinical efficacy of these kinase-targeted drugs including the challenges of tumor heterogeneity and microenvironment as well as the emergence of mutations that confer drug resistance. Another major challenge of kinase inhibition is that of the development of adverse side effects. Some classic examples of this include dermatologic complications and cardiotoxicity associated with inhibition of EGFR and vascular endothelial growth factor receptor (VEGFR), respectively. Furthermore, there is an urgent need to develop relevant models of resistance in response to kinase inhibitors in efforts to overcome this resistance via potential synergistic combinatorial therapies.

Another critical issue facing clinical trial design with kinase-targeted agents is that of determining the types of tumors that are most likely to respond to specific kinase inhibitors and thus identify the subsets of patients who will likely benefit from these treatments. To combat this issue, many studies have been dedicated to identifying certain "kinase dependencies" in cancer cells that would make them more susceptible to inhibition. These so-called dependencies are primarily based on the existence of constitutively activate kinases achieved by gene mutation, amplification, or fusion. Among the potential approaches to identifying signatures of kinase dependency are proteomic profiling, next-generation sequencing and various applications utilizing phospho-specific antibodies against numerous specific kinase substrates. Additional mechanisms of kinase dependency include impairment of the function of phosphatases, the negative regulators of phosphorylation as is the case with mutations in the phosphatase and tensin homolog (PTEN) tumor suppressor gene. The consequence of *PTEN* loss is signal propagation through downstream kinases such as Akt. Moreover, growing evidence from isogenic human and mouse models also suggests that this type of indirect avenue of kinase dependency may be analogous to direct, activating mutations in the kinases themselves.

Finally, there are also cases in which the beneficial effect of a kinase inhibitor is counteracted by an additional genetic lesion in a compensatory signaling pathway. Therefore, studies to identify such secondary events are urgently needed. Taken together, these evidences underscore the critical need to optimize the use of kinase inhibitors against cancers by continued detailed molecular characterization of tumor tissues. It will be critical to develop new compounds that circumvent acquired resistance to the first-generation kinase inhibitors for patients with refractory disease.

3.3 Cutting edge therapeutics for treating cancers with constitutively active kinases

Since the mid-1970s, numerous studies have highlighted a crucial role for kinases in promoting tumorigenesis and metastasis [63–67]. This is of no surprise, since a clear majority of protein kinases promote critical cellular functions pertinent to cancer progression, including cell proliferation, survival, and migration [68–72]. Hence, targeting mutated, overexpressed, or hyperactive kinases represents an important and promising clinical niche for developing drugs for cancer therapy [67, 73, 74]. Interestingly, inhibitors against kinases account for approximately 25% of all current drug discovery efforts. So far, ~37 inhibitors with activity targeted to one or multiple kinases have been approved for clinical use [67, 75]. These range from highly selective monoclonal antibodies to more broad-spectrum synthetic or natural small molecules that have achieved a significant increase in patient survival rate in cancers [67]. For example, treatment with imatinib and dasatinib, which are highly potent and selective tyrosine kinase inhibitors against BCR-ABL, produces more favorable outcomes compared to conventional cytotoxic therapy for patients with chronic myeloid leukemia (CML) [76–78]. Similarly, broad-spectrum inhibitors have also been used with great success. For example, CHIR-258, which targets multiple kinases, inhibits VEGFR, platelet-derived growth factor receptor (PDGFR), FMS-like tyrosine kinase-3 (FLT-3), mast/stem cell growth factor receptor (c-Kit), and fibroblast growth factor receptor (FGFR) in multiple myeloma patients and is most effective for killing tumors harboring the translocation t [4, 14] (p16.3;q32.3) with increased expression and activating mutations of FGFR3 [79–81].

Some other major inhibitors to enter the market include those targeting key oncogenic kinase drug targets such as ERBB2 (e.g., afatinib), HER2 (e.g., trastuzumab), and VEGFRs (e.g., sorafenib, a small molecule inhibitor used for the treatment of renal cell, liver. and thyroid cancers) [82–87]. Another successful targeting strategy has been the use of a monoclonal antibody against VEGFR (bevacizumab). Bevacizumab is used in combination with chemotherapy to treat patients with metastatic colon cancer, and its widespread use has resulted in significant improvement in survival outcomes [88]. Other examples include inhibitors against Kit, PDGFRs, proto-oncogene tyrosine-protein kinase Src (SRC), mechanistic target of rapamycin and FK506-binding protein 12-rapamycin-associated protein 1 (mTOR), phosphatidylinositol-4,5-bisphosphate 3-kinase catalytic subunit alpha (PIK3CA), serine/threonine-protein kinase B (PKB or Akt)—Raf (BRAF), and epidermal growth factor receptor (EGFR) (e.g., cetuximab, panitumumab), all of which act to activate significant tumor cell signaling pathways such as NF-κB [67, 89]. For instance, several FDA-approved kinase inhibitors, although not perceived as direct NF-κB inhibitors, have been shown to suppress NF-κB signaling [89]. These include inhibitors targeting EGFR in breast and lung cancer, Akt in breast cancer, GSK-3β in pancreatic cancer, NIK in melanoma, BRAF in multiple myeloma, and IKK in brain and liver cancer [89–91]. Specifically, GSK2118436, PLX4720, sorafenib, and PLX4032 are all drugs which are currently being used to target B-Raf^{V600E} in advanced cancers with elevated NF-κB activity [73, 92–94]. Finally, small molecule inhibitors targeting Akt, which include perifosine, GSK690693, VQD002, and MK2206, are also being tested clinically [95, 96].

In summary, these studies highlight the importance of inhibition of distinct kinase signaling pathways as a means of minimizing cytotoxic effects on non-cancerous cells, thus bestowing selective killing of tumor cells and improving patient clinical outcomes.

4. Conclusion and future directions

In summary, this chapter highlights a significant role of NF-κB phosphorylation in driving the initiation and progression of several cancers as well as chemoresistance to first-line therapies. Furthermore, we emphasized the relationship between p65 phosphorylation and the role that constitutively active kinases play in promoting the cancer phenotype, via utilization of ATP as the phosphate donor. Finally, we underlined the well-established therapeutic potential of targeting these kinases in the treatment of various cancers. Despite these encouraging data, we acknowledge that the difficulties with drug resistance and toxicity continue to present critical challenges for the use of kinase inhibitors in both clinical and experimental oncology. Furthermore, issues related to the inadequate understanding of the selectivity of the kinase inhibitors have also plagued the successful clinical utility of these inhibitors. Nevertheless, a key challenge for overcoming this enigma is to identify the most efficacious, complementary, and least toxic combinations of kinase inhibitors for targeted cancer treatment [97, 98]. This will likely lead to the development of multimodal treatment initiatives that evade the treatment-related drug resistance. Finally, it is well known that cancers have heterogeneous populations of cells, and these may differentially contribute to chemoresistance [99, 100]. To address this, many efforts are underway to eliminate cancer stem cells as main culprits of this intrinsic heterogeneity, which will undoubtedly improve our understanding of better drug design and efficiency [101, 102].

Acknowledgements

This publication is made possible, in part, with support from the Indiana Clinical and Translational Sciences Institute (CTSI) funded from the National Institutes of Health, National Center for Advancing Translational Sciences, Clinical and Translational Sciences Award (to TL); V foundation Kay Yow Cancer Fund (Grant 4486242 to TL); NIH-NIGMS Grant (#1R01GM120156-01A1 (to TL); and 100 VOH Grant (#2987613 to TL), as well as NIH-NCI Grant (#1R03CA223906-01).

Author details

Matthew Martin[1], Antja-Voy Hartley[1], Jiamin Jin[1], Mengyao Sun[1] and Tao Lu[1,2,3]*

1 Department of Pharmacology and Toxicology, Indiana University School of Medicine, USA

2 Department of Biochemistry and Molecular Biology, Indiana University School of Medicine, USA

3 Department of Medical and Molecular Genetics, Indiana University School of Medicine, USA

*Address all correspondence to: lut@iu.edu

References

[1] Jost PJ, Ruland J. Aberrant NF-κB signaling in lymphoma: Mechanisms, consequences, and therapeutic implications. Blood. 2007;**109**(7):2700-2707

[2] Nogueira L et al. The NFκB pathway: A therapeutic target in glioblastoma. Oncotarget. 2011;**2**(8):646

[3] Sen R, Baltimore D. Multiple nuclear factors interact with the immunoglobulin enhancer sequences. Cell. 1986;**46**(5):705-716

[4] Lu T, Stark GR. NF-κB: Regulation by methylation. Cancer Research. 2015;**75**(18):3692-3695

[5] Vaiopoulos AG, Athanasoula KC, Papavassiliou AG. NF-κB in colorectal cancer. Journal of Molecular Medicine. 2013;**91**(9):1029-1037

[6] Xia Y, Shen S, Verma IM. NF-κB, an active player in human cancers. Cancer Immunology Research. 2014;**2**(9):823-830

[7] Pacifico F, Leonardi A. Role of NF-κB in thyroid cancer. Molecular and Cellular Endocrinology. 2010;**321**(1):29-35

[8] Morais C et al. The emerging role of nuclear factor kappa B in renal cell carcinoma. The International Journal of Biochemistry & Cell Biology. 2011;**43**(11):1537-1549

[9] Martin M et al. Novel serine 176 phosphorylation of YBX1 activates NF-κB in colon cancer. Journal of Biological Chemistry. 2017. jbc-M116

[10] Prabhu L et al. Critical role of phosphorylation of serine 165 of YBX1 on the activation of NF-κB in colon cancer. Oncotarget. 2015;**6**(30):29396

[11] Prabhu L et al. Critical role of NF-κB in pancreatic cancer. Oncotarget. 2014;**5**(22):10969

[12] Smyth MJ, Dunn GP, Schreiber RD. Cancer immunosurveillance and immunoediting: The roles of immunity in suppressing tumor development and shaping tumor immunogenicity. Advances in Immunology. 2006;**90**:1-50

[13] Hinz M et al. NF-κB function in growth control: Regulation of cyclin D1 expression and G0/G1-to-S-phase transition. Molecular and Cellular Biology. 1999;**19**(4):2690-2698

[14] Wu Y et al. Stabilization of snail by NF-κB is required for inflammation-induced cell migration and invasion. Cancer Cell. 2009;**15**(5):416-428

[15] Gasparini C et al. NF-κB pathways in hematological malignancies. Cellular and Molecular Life Sciences. 2014;**71**(11):2083-2102

[16] Viatour P et al. Phosphorylation of NF-κB and IκB proteins: Implications in cancer and inflammation. Trends in Biochemical Sciences. 2005;**30**(1):43-52

[17] Hu J et al. Insufficient p65 phosphorylation at S536 specifically contributes to the lack of NF-κB activation and transformation in resistant JB6 cells. Carcinogenesis. 2004;**25**(10):1991-2003

[18] Wang Y-P et al. Astrocyte elevated gene-1 is associated with metastasis in head and neck squamous cell carcinoma through p65 phosphorylation and upregulation of MMP1. Molecular Cancer. 2013;**12**(1):109

[19] Zhang L et al. Function of phosphorylation of NF-kB p65 ser536 in prostate cancer oncogenesis. Oncotarget. 2015;**6**(8):6281

[20] Bonnard M et al. Deficiency of T2K leads to apoptotic liver degeneration and impaired NF-κB-dependent gene transcription. The EMBO Journal. 2000;**19**(18):4976-4985

[21] Hoeflich KP et al. Requirement for glycogen synthase kinase-3β in cell survival and NF-κB activation. Nature. 2000;**406**(6791):86

[22] Schwabe RF, Brenner DA. Role of glycogen synthase kinase-3 in TNF-α-induced NF-κB activation and apoptosis in hepatocytes. American Journal of Physiology-Gastrointestinal and Liver Physiology. 2002;**283**(1):G204-G211

[23] Fujita F et al. Identification of NAP1, a regulatory subunit of IκB kinase-related kinases that potentiates NF-κB signaling. Molecular and Cellular Biology. 2003;**23**(21):7780-7793

[24] Ly Philip TT et al. Inhibition of GSK3β-mediated BACE1 expression reduces Alzheimer-associated phenotypes. The Journal of Clinical Investigation. 2012;**123**(1)

[25] Zhong H et al. The transcriptional activity of NF-κB is regulated by the IκB-associated PKAc subunit through a cyclic AMP–independent mechanism. Cell. 1997;**89**(3):413-424

[26] Zhong H, Voll RE, Ghosh S. Phosphorylation of NF-κB p65 by PKA stimulates transcriptional activity by promoting a novel bivalent interaction with the coactivator CBP/p300. Molecular Cell. 1998;**1**(5):661-671

[27] Zhong H et al. The phosphorylation status of nuclear NF-κB determines its association with CBP/p300 or HDAC-1. Molecular Cell. 2002;**9**(3):625-636

[28] Vermeulen L et al. Transcriptional activation of the NF-κB p65 subunit by mitogen-and stress-activated protein kinase-1 (MSK1). The EMBO Journal. 2003;**22**(6):1313-1324

[29] Nihira K et al. Pim-1 controls NF-κB signalling by stabilizing RelA/p65. Cell Death and Differentiation. 2010;**17**(4):689

[30] Wang H et al. Proteinase-activated receptors induce interleukin-8 expression by intestinal epithelial cells through ERK/RSK90 activation and histone acetylation. The FASEB Journal. 2010;**24**(6):1971-1980

[31] Wang Y et al. M-CSF induces monocyte survival by activating NF-κB p65 phosphorylation at Ser276 via protein kinase C. PLoS One. 2011;**6**(12):e28081

[32] Duncan JS, Litchfield DW. Too much of a good thing: the role of protein kinase CK2 in tumorigenesis and prospects for therapeutic inhibition of CK2. Biochimica et Biophysica Acta (BBA)-Proteins and Proteomics. 2008;**1784**(1):33-47

[33] Wang D et al. TNFalpha-induced phosphorylation of RelA/p65 on Ser529 is controlled by casein kinase II. Journal of Biological Chemistry. 2000

[34] Arun P et al. Nuclear NF-κB p65 phosphorylation at serine 276 by protein kinase A contributes to the malignant phenotype of head and neck cancer. Clinical Cancer Research. 2009;**15**(19):5974-5984

[35] Buss H et al. Cyclin-dependent kinase 6 phosphorylates NF-κB P65 at serine 536 and contributes to the regulation of inflammatory gene expression. PLoS One. 2012;**7**(12):e51847

[36] Valdman A et al. Pim-1 expression in prostatic intraepithelial neoplasia and human prostate cancer. The Prostate. 2004;**60**(4):367-371

[37] Lam YP, di Tomaso E, Ng H-K, Pang JCS, Roussel MF, Hjelm NM. Expression of p19 INK4d, CDK4, CDK6 in

glioblastoma multiforme. British Journal of Neurosurgery. 2000;**14**(1):28-32

[38] Mendrzyk F et al. Genomic and protein expression profiling identifies CDK6 as novel independent prognostic marker in medulloblastoma. Journal of Clinical Oncology. 2005;**23**(34):8853-8862

[39] Igarashi K et al. Activation of cyclin D1-related kinase in human lung adenocarcinoma. British Journal of Cancer. 1999;**81**(4):705

[40] Katsha A et al. Aurora kinase A promotes inflammation and tumorigenesis in mice and human gastric neoplasia. Gastroenterology. 2013;**145**(6):1312-1322

[41] Wang B et al. Role of novel serine 316 phosphorylation of the p65 subunit of NF-κB in differential gene regulation. Journal of Biological Chemistry. 2015;jbc-M115

[42] Pan X et al. Nuclear factor-κB p65/relA silencing induces apoptosis and increases gemcitabine effectiveness in a subset of pancreatic cancer cells. Clinical Cancer Research. 2008;**14**(24):8143-8151

[43] Zemskova M et al. The PIM1 kinase is a critical component of a survival pathway activated by docetaxel and promotes survival of docetaxel-treated prostate cancer cells. Journal of Biological Chemistry. 2008;**283**(30):20635-20644

[44] Uetsuka H et al. Inhibition of inducible NF-κB activity reduces chemoresistance to 5-fluorouracil in human stomach cancer cell line. Experimental Cell Research. 2003;**289**(1):27-35

[45] Li F, Gautam S. Targeting transcription factor NF-κB to overcome chemoresistance and radioresistance in cancer therapy. Biochimica et Biophysica Acta (BBA)-Reviews on Cancer. 2010;**1805**(2):167-180

[46] Godwin P et al. Targeting nuclear factor-kappa B to overcome resistance to chemotherapy. Frontiers in Oncology. 2013;**3**:120

[47] Bednarski BK et al. Active roles for inhibitory κB kinases α and β in nuclear factor-κB–mediated chemoresistance to doxorubicin. Molecular Cancer Therapeutics. 2008;**7**(7):1827-1835

[48] Guo J et al. Enhanced chemosensitivity to irinotecan by RNA interference-mediated down-regulation of the nuclear factor-κB p65 subunit. Clinical Cancer Research. 2004;**10**(10):3333-3341

[49] Wang W, McLeod HL, Cassidy J. Disulfiram-mediated inhibition of NF-κB activity enhances cytotoxicity of 5-fluorouracil in human colorectal cancer cell lines. International Journal of Cancer. 2003;**104**(4):504-511

[50] Baumann P et al. Alkylating agents induce activation of NFκB in multiple myeloma cells. Leukemia Research. 2008;**32**(7):1144-1147

[51] Dai Y et al. Blockade of histone deacetylase inhibitor-induced RelA/p65 acetylation and NF-κB activation potentiates apoptosis in leukemia cells through a process mediated by oxidative damage, XIAP downregulation, and c-Jun N-terminal kinase 1 activation. Molecular and Cellular Biology. 2005;**25**(13):5429-5444

[52] Wu Z-H et al. Molecular linkage between the kinase ATM and NF-κB signaling in response to genotoxic stimuli. Science. 2006;**311**(5764):1141-1146

[53] Cusack JC et al. Enhanced chemosensitivity to CPT-11 with proteasome inhibitor PS-341: Implications for systemic nuclear

factor-κB inhibition. Cancer Research. 2001;**61**(9):3535-3540

[54] Denlinger CE et al. Combined proteasome and histone deacetylase inhibition in non–small cell lung cancer. The Journal of Thoracic and Cardiovascular Surgery. 2004;**127**(4):1078-1086

[55] Ravaud A et al. Phase I study and pharmacokinetic of CHS-828, a guanidino-containing compound, administered orally as a single dose every 3 weeks in solid tumours: An ECSG/EORTC study. European Journal of Cancer. 2005;**41**(5):702-707

[56] Huang J-J et al. Novel IKKβ inhibitors discovery based on the co-crystal structure by using binding-conformation-based and ligand-based method. European Journal of Medicinal Chemistry. 2013;**63**:269-278

[57] Matsumoto R et al. Inhibition of IκB phosphorylation by a novel IKK inhibitor IMD-1041 attenuates myocardial dysfunction after infarction. Immunology, Endocrine & Metabolic Agents in Medicinal Chemistry. 2012;**12**(2):137-142

[58] Levidou G et al. Expression of nuclear factor κB in human gastric carcinoma: Relationship with IκBa and prognostic significance. Virchows Archiv. 2007;**450**(5):519-527

[59] Ougolkov AV et al. Glycogen synthase kinase-3β participates in nuclear factor κB–mediated gene transcription and cell survival in pancreatic cancer cells. Cancer Research. 2005;**65**(6):2076-2081

[60] Shishodia S et al. Ursolic acid inhibits nuclear factor-κB activation induced by carcinogenic agents through suppression of IκBα kinase and p65 phosphorylation: Correlation with down-regulation of cyclooxygenase 2, matrix metalloproteinase 9, and cyclin D1. Cancer Research. 2003;**63**(15):4375-4383

[61] Natarajan K et al. Caffeic acid phenethyl ester is a potent and specific inhibitor of activation of nuclear transcription factor NF-kappa B. Proceedings of the National Academy of Sciences. 1996;**93**(17):9090-9095

[62] Finco TS, Beg AA, Baldwin AS. Inducible phosphorylation of I kappa B alpha is not sufficient for its dissociation from NF-kappa B and is inhibited by protease inhibitors. Proceedings of the National Academy of Sciences. 1994;**91**(25):11884-11888

[63] Drake JM, Lee JK, Witte ON. Clinical targeting of mutated and wild-type protein tyrosine kinases in cancer. Molecular and Cellular Biology. 2014;**MCB-01592**

[64] Gross S et al. Targeting cancer with kinase inhibitors. The Journal of Clinical Investigation. 2015;**125**(5):1780-1789

[65] Paul MK, Mukhopadhyay AK. Tyrosine kinase—Role and significance in cancer. International Journal of Medical Sciences. 2004;**1**(2):101

[66] Shchemelinin I, Sefc L, Necas E. Protein kinases, their function and implication in cancer and other diseases. Folia Biologica. 2006;**52**(3):81

[67] Bhullar KS et al. Kinase-targeted cancer therapies: Progress, challenges and future directions. Molecular Cancer. 2018;**17**(1):48

[68] Lemmon MA, Schlessinger J. Cell signaling by receptor tyrosine kinases. Cell. 2010;**141**(7):1117-1134

[69] Kaistha BP et al. Key role of dual specificity kinase TTK in proliferation and survival of pancreatic cancer cells. British Journal of Cancer. 2014;**111**(9):1780

[70] Duncan JS et al. Regulation of cell proliferation and survival: Convergence of protein kinases and caspases. Biochimica et Biophysica Acta (BBA): Proteins and Proteomics. 2010;**1804**(3):505-510

[71] Cain RJ, Ridley AJ. Phosphoinositide 3-kinases in cell migration. Biology of the Cell. 2009;**101**(1):13-29

[72] Huang C, Jacobson K, Schaller MD. MAP kinases and cell migration. Journal of Cell Science. 2004;**117**(20):4619-4628

[73] Regad T. Targeting RTK signaling pathways in cancer. Cancers. 2015;7(3):1758-1784

[74] Maurer G, Tarkowski B, Baccarini M. Raf kinases in cancer—Roles and therapeutic opportunities. Oncogene. 2011;**30**(32):3477

[75] Xu J et al. Comparison of FDA approved kinase targets to clinical trial ones: Insights from their system profiles and drug-target interaction networks. BioMed Research International. 2016

[76] Steinberg M. Dasatinib: A tyrosine kinase inhibitor for the treatment of chronic myelogenous leukemia and Philadelphia chromosome—Positive acute lymphoblastic leukemia. Clinical Therapeutics. 2007;**29**(11):2289-2308

[77] Kantarjian H et al. Nilotinib in imatinib-resistant CML and Philadelphia chromosome–positive ALL. New England Journal of Medicine. 2006;**354**(24):2542-2551

[78] Schultz KR et al. Improved early event-free survival with imatinib in Philadelphia chromosome–positive acute lymphoblastic leukemia: A children's oncology group study. Journal of Clinical Oncology. 2009;**27**(31):5175

[79] Chesi M et al. Frequent translocation t (4; 14)(p16. 3; q32. 3) in multiple myeloma is associated with increased expression and activating mutations of fibroblast growth factor receptor 3. Nature Genetics. 1997;**16**(3):260

[80] De Menezes, Lopes DE, et al. CHIR-258: A potent inhibitor of FLT3 kinase in experimental tumor xenograft models of human acute myelogenous leukemia. Clinical Cancer Research. 2005;**11**(14):5281-5291

[81] Trudel S et al. CHIR-258, a novel, multitargeted tyrosine kinase inhibitor for the potential treatment of t (4; 14) multiple myeloma. Blood. 2005;**105**(7):2941-2948

[82] Fabian MA et al. A small molecule–kinase interaction map for clinical kinase inhibitors. Nature Biotechnology. 2005;**23**(3):329

[83] Yang JC-H et al. Afatinib versus cisplatin-based chemotherapy for EGFR mutation-positive lung adenocarcinoma (LUX-lung 3 and LUX-lung 6): Analysis of overall survival data from two randomised, phase 3 trials. The Lancet Oncology. 2015;**16**(2):141-151

[84] Soria J-C et al. Afatinib versus erlotinib as second-line treatment of patients with advanced squamous cell carcinoma of the lung (LUX-lung 8): An open-label randomised controlled phase 3 trial. The Lancet Oncology. 2015;**16**(8):897-907

[85] Kato T et al. Afatinib versus cisplatin plus pemetrexed in Japanese patients with advanced non-small cell lung cancer harboring activating EGFR mutations: Subgroup analysis of LUX-lung 3. Cancer Science. 2015;**106**(9):1202-1211

[86] Jackisch C et al. Subcutaneous versus intravenous formulation of trastuzumab for HER2-positive early breast cancer: Updated results from the phase III HannaH study. Annals of Oncology. 2014;**26**(2):320-325

[87] Pitoia F, Jerkovich F. Selective use of sorafenib in the treatment of thyroid cancer. Drug Design, Development and Therapy. 2016;**10**:1119

[88] Ilic I, Jankovic S, Ilic M. Bevacizumab combined with chemotherapy improves survival for patients with metastatic colorectal cancer: Evidence from meta analysis. PLoS One. 2016;**11**(8):e0161912

[89] Chaturvedi MM et al. NF-κB addiction and its role in cancer:'one size does not fit all'. Oncogene. 2011;**30**(14):1615

[90] Pianetti S et al. Her-2/neu overexpression induces NF-κB via a PI3-kinase/Akt pathway involving calpain-mediated degradation of IκB-α that can be inhibited by the tumor suppressor PTEN. Oncogene. 2001;**20**(11):1287

[91] Keats JJ et al. Promiscuous mutations activate the noncanonical NF-κB pathway in multiple myeloma. Cancer Cell. 2007;**12**(2):131-144

[92] Lin K et al. The role of B-RAF mutations in melanoma and the induction of EMT via dysregulation of the NF-κB/snail/RKIP/PTEN circuit. Genes & Cancer. 2010;**1**(5):409-420

[93] Li W et al. The relationship between BRAFV600E, NF-κB and TgAb expression in papillary thyroid carcinoma. Pathology-Research and Practice. 2017;**213**(3):183-188

[94] Fuchs O. Targeting of NF-kappaB signaling pathway, other signaling pathways and epigenetics in therapy of multiple myeloma. Cardiovascular & Haematological Disorders-Drug Targets. 2013;**13**(1):16-34

[95] Mattmann ME, Stoops SL, Lindsley CW. Inhibition of Akt with small molecules and biologics: Historical perspective and current status of the patent landscape. Expert Opinion on Therapeutic Patents. 2011;**21**(9):1309-1338

[96] Hussain AR et al. Cross-talk between NFkB and the PI3-kinase/AKT pathway can be targeted in primary effusion lymphoma (PEL) cell lines for efficient apoptosis. PLoS One. 2012;**7**(6):e39945

[97] Saha D et al. Combinatorial effects of VEGFR kinase inhibitor axitinib and oncolytic virotherapy in mouse and human glioblastoma stem-like cell models. Clinical Cancer Research. 2018;clincanres-1717

[98] Ahronian LG, Corcoran RB. Strategies for monitoring and combating resistance to combination kinase inhibitors for cancer therapy. Genome Medicine. 2017;**9**(1):37

[99] Asim BM et al. Intratumoral heterogeneity and chemoresistance in nonseminomatous germ cell tumor of the testis. Oncotarget. 2016;7(52):86280

[100] Brooks MD, Burness ML, Wicha MS. Therapeutic implications of cellular heterogeneity and plasticity in breast cancer. Cell Stem Cell. 2015;**17**(3):260-271

[101] Prasetyanti PR, Medema JP. Intra-tumor heterogeneity from a cancer stem cell perspective. Molecular Cancer. 2017;**16**(1):41

[102] Dragu DL et al. Therapies targeting cancer stem cells: Current trends and future challenges. World Journal of Stem Cells. 2015;7(9):1185

[103] Hoesel B, Schmid JA. The complexity of NF-κB signaling in inflammation and cancer. Molecular Cancer. 2013;**12**(1):86

[104] Han X et al. Phosphorylation of mini-chromosome maintenance 3 (MCM3) by Chk1 negatively regulates DNA replication and checkpoint

activation. Journal of Biological Chemistry. 2015:jbc-M114

[105] Ryo A et al. Regulation of NF-κB signaling by Pin1-dependent prolyl isomerization and ubiquitin-mediated proteolysis of p65/RelA. Molecular Cell. 2003;**12**(6):1413-1426

[106] Anrather J, Racchumi G, Iadecola C. Cis-acting element-specific transcriptional activity of differentially phosphorylated nuclear factor-κB. Journal of Biological Chemistry. 2005;**280**(1):244-252

[107] Duran A, Diaz-Meco MT, Moscat J. Essential role of RelA Ser311 phosphorylation by ζPKC in NF-κB transcriptional activation. The EMBO Journal. 2003;**22**(15):3910-3918

[108] Yeh PY et al. Suppression of MEK/ERK signaling pathway enhances cisplatin-induced NF-κB activation by protein phosphatase 4-mediated NF-κB p65 Thr dephosphorylation. Journal of Biological Chemistry. 2004;**279**(25):26143-26148

[109] Buss H et al. Phosphorylation of serine 468 by GSK-3β negatively regulates basal p65 NF-κB activity. Journal of Biological Chemistry. 2004;**279**(48):49571-49574

[110] Rocha S et al. Regulation of NF-κB and p53 through activation of ATR and Chk1 by the ARF tumour suppressor. The EMBO Journal. 2005;**24**(6):1157-1169

[111] Bae JS et al. Phosphorylation of NF-κB by calmodulin-dependent kinase IV activates anti-apoptotic gene expression. Biochemical and Biophysical Research Communications. 2003;**305**(4):1094-1098

[112] Sabatel H et al. Phosphorylation of p65 (RelA) on Ser547 by ATM represses NF-κB-dependent transcription of specific genes after genotoxic stress. PLoS One. 2012;**7**(6):e38246

Chapter 5

Clinical Evaluation of Adenosine Triphosphate Disodium Hydrate (ATP-2Na) for Asthenopia

Yo Nakamura, Yukiko Ban, Yoko Ikeda, Hisayo Higashihara and Shigeru Kinoshita

Abstract

To investigate the effect and the safety of Adenosine triphosphate disodium hydrate (ATP-2Na) for asthenopia. 40 subjects [35 females and 5 males, 25~87 years old (average: 62.5 years old)] with asthenopia ingested 200~300 mg/day ATP-2Na for 3 months. Before and after 1 and 3 months ingestion, subjects completed a questionnaire to determine their asthenopia symptom and fatigue symptom by visual analog scale (VAS). The scores were compared between before and after ingestion. 31 subjects completed a questionnaire for 1 month. The scores of asthenopia symptom before ingestion, 1 and 3 months were 4.05 ± 3.22, 2.67 ± 2.19 and 2.41 ± 2.16, respectively. The scores of fatigue symptom were 4.76 ± 3.05, 3.08 ± 2.93 and 3.10 ± 3.19, respectively. Both scores were significantly decreased ($p < 0.005$) at 1 month compared before ingestion. Three subjects had side effects (diarrhea for two, nausea for one), and all subjects improved by oral discontinuation. These results suggest that ATP-2Na is relatively early effective in improving asthenopia and accompanying fatigue symptoms.

Keywords: adenosine triphosphate disodium hydrate(ATP-2Na), asthenopia symptom, fatigue symptom, visual analog scale (VAS), 3 months

1. Introduction

Recent years, patients with asthenopia increased, because many people are engaged in visual display terminal (VDT) work and are using a portable terminal for example smartphone. The causes of asthenopia are dry eye [1, 2], lack of correction for hyperopia or presbyopia, overcorrection for myopia by contact lens or glasses for example. Eye treatment for these causes does not improve asthenopia in some cases. On the other hand, patients with asthenopia sometimes complain of systemic symptoms such as headache, stiff shoulder, nausea and fatigue. Symptoms by VDT work are serious problems in Japan, so Ministry of Health, Labor and Welfare established guideline for Occupational Health Environmental Management in 2002 [3]. In this guideline, at first, improvement of environment for VDT is necessary, and in serious cases patients should see clinical ophthalmologists. In questionnaire results in 2003 in Japan, many ophthalmologists prescribe glasses for near work or tear drop or Vitamin B12 eye drop to reduce asthenopia from VDT work [4].

Adenosine triphosphate disodium hydrate (ATP-2Na) is internal medicine which approved for asthenopia from the results of multicenter clinical researches or double blind clinical researches in 1970~1990 in Japan [5–7]. In this study, we researched the effect and the safety of ATP-2Na for present asthenopia.

2. Subjects and methods

2.1 Subjects

Subjects were patients with asthenopia in Kyoto Prefectural University of Medicine facilities in 2012~2014. We selected patients who had asthenopia symptoms, regardless of causes such as dry eye or inappropriate glasses. Subjects were 40 patients (35 females and 5 males), and 25~87 years old (average age: 62.5 years old).

2.2 Methods

Two or three times per day, subjects ingested Adetphos Kowa Granule 10% which is clinical medicine containing 100 mg of ATP-2Na. Before and after 1 and 3

		Date / /
Eye symptoms	Feeling of oppression of the eyes	0 (Non) ——— 10 (Worst)
	Pain in the back of the eyes	0 (Non) ——— 10 (Worst)
	Tough to open eyes	0 (Non) ——— 10 (Worst)
	Blurred vision	0 (Non) ——— 10 (Worst)
	Sense of heat of eyes	0 (Non) ——— 10 (Worst)
Systemic symptoms	Headache	0 (Non) ——— 10 (Worst)
	Stiff shoulder	0 (Non) ——— 10 (Worst)
	Fatigue	0 (Non) ——— 10 (Worst)
Mental symptoms	Depressed feeling	0 (Non) ——— 10 (Worst)
	No concentration feeling	0 (Non) ——— 10 (Worst)
Remarks column		

Table 1.
Visual analog scale (VAS).

months ingestion, subjects completed a questionnaire to determine their asthenopia symptom and fatigue symptom by visual analog scale (VAS) [8] (**Table 1**). In VAS scale, 0 is the best evaluated result and 10 is the worst evaluated result. Questionnaire items were 10 items; 5 items about eye symptoms (feeling of oppression of the eyes, pain in the back of the eyes, tough to open eyes, blurred vision, sense of heat of eyes), 3 items about systemic symptoms (headache, stiff shoulder, fatigue), and 2 items about mental symptoms (depressed feeling, no concentration feeling). The scores were compared between before and after ingestion.

In addition, the scores were compared about the background factors before administration. Researched background factors were gender, age, internal dose per body weight, whether or not other eye diseases, whether or not of using eye drop for asthenopia, and scores before ingestion. Within this study, subjects continued all drugs prescribed before the administration.

2.3 Statistical analysis

Paired t test was used to compare the change of scores between before and after ingestion. In analysis of background factors, student's t test was used the changes through 3 months. In both analysis, $p < 0.05$ was considered statistically significant.

3. Results

3.1 Subject characteristics

In 40 initial subjects, 33 subjects visited hospital at 1 month, and these subjects were targeted for evaluation of safety. In 33 subjects, one subject discontinued the ingestion because of diarrhea within 1 month, one refused to describe the questionnaire. 31 subjects completed a questionnaire for 1 month and were targeted for effect evaluation. Age distribution of 31 subjects (26 females, 5 males) was 25–87 years old, average age was 64.4 years old. Background factors are presented in **Table 2**. Average score of eye symptom in group with diabetes was larger than that in group without diabetes ($p < 0.01$). Average score of systemic symptom in group 65y younger was larger than that in group 65y older ($p < 0.05$). 26 subjects completed a questionnaire for 3 months.

3.2 Effect evaluations

3.2.1 Scores in each symptom

The scores about eye symptom before ingestion, 1 month, 3 months were 4.05 ± 3.22, 2.67 ± 2.19, and 2.41 ± 2.16, respectively. The scores were significantly decreased at 1 and 3 months (both $p < 0.005$, **Figure 1**). Among the symptoms about eye, the scores of "blurred vision" were no different between before and after (**Figure 2**). The scores about systemic symptom (average scores; before ingestion: 4.76 ± 3.05, 1 month: 3.08 ± 2.93, and 3 months: 3.13 ± 3.18) significantly decreased ($p < 0.005$). Similarly, the scores about mental symptom (average scores; before ingestion: 4.41 ± 3.46, 1 month: 2.97 ± 2.63, and 3 months: 3.10 ± 3.19) significantly decreased ($p < 0.005$).

		Score of first time		P
Sex		Female(n=26)	Male(n=5)	
	Eye symptoms	4.33±2.45	2.44±1.29	p=0.11
	Systemic symptoms	4.93±2.97	3.87±3.69	p=0.49
	Mental symptoms	4.51±3.53	3.90±3.35	p=0.72
Age		Under 65 years old(n=12)	Over 65 years old(n=19)	
	Eye symptoms	4.31±1.89	3.84±2.70	p=0.61
	Systemic symptoms	6.36±2.60	3.75±2.94	p<0.05
	Mental symptoms	4.99±3.41	4.04±3.53	p=0.47
Involvement				
Cataract		Yes(n=12)	No(n=19)	
	Eye symptoms	4.58±2.79	3.67±2.11	p=0.31
	Systemic symptoms	4.73±3.00	4.78±3.17	p=0.96
	Mental symptoms	5.40±3.45	3.78±3.40	p=0.21
Glaucoma		Yes(n=11)	No(n=20)	
	Eye symptoms	4.90±2.81	3.54±2.05	p=0.13
	Systemic symptoms	5.72±3.24	4.23±2.89	p=0.20
	Mental symptoms	5.05±3.72	4.05±3.35	p=0.45
Hypertension		Yes(n=11)	No(n=20)	
	Eye symptoms	3.62±2.74	4.24±2.22	p=0.50
	Systemic symptoms	4.27±3.25	5.03±2.99	p=0.52
	Mental symptoms	3.86±3.52	4.71±3.47	p=0.52
Diabetes		Yes(n=3)	No(n=28)	
	Eye symptoms	7.30±2.22	3.67±2.16	p<0.01
	Systemic symptoms	5.41±3.84	4.69±3.04	p=0.71
	Mental symptoms	4.88±3.87	4.36±3.48	p=0.81
Dose		200mg (n=23)	300mg(n=8)	
	Eye symptoms	4.31±2.57	3.20±1.65	p=0.27
	Systemic symptoms	5.09±3.09	3.82±2.93	p=0.32
	Mental symptoms	4.94±3.44	2.88±3.21	p=0.15
Other eye strain treatments		Yes(n=13)※1	No(n=18)	
	Eye symptoms	4.55±2.65	3.64±2.19	p=0.30
	Systemic symptoms	4.83±3.16	4.71±3.07	p=0.92
	Mental symptoms	4.27±3.70	4.51±3.38	p=0.85

VAS: Visual Analog Scale

※1 Concomitant medications: Oral administration of Methylcobalamin (6 person), Instillation of cyanocobalamin (4 person), Oral administration of Methylcobalamin + Instillation of cyanocobalamin (1 person), Over-the-counter drugs (2 person)

Table 2.
Comparison of baseline VAS scores between each groups classified by baseline characteristics (n = 31).

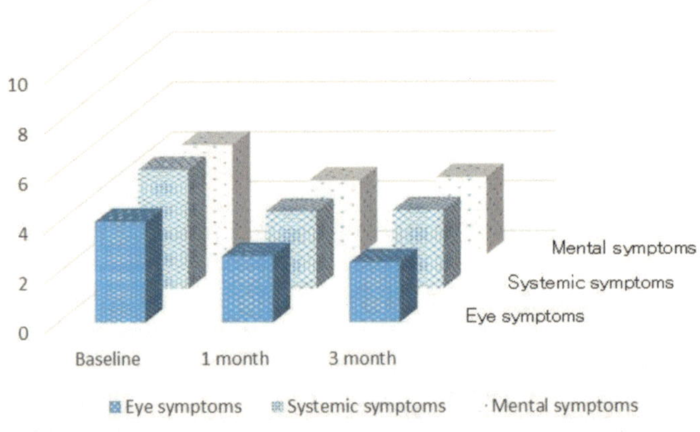

Figure 1.
Transition of scores of each symptom. Mean score of each symptom (5 eye symptoms, 3 systemic symptoms, 2 mental symptoms) before and 1, 3 months after ingestion. The scores were significantly decreased at 1 and 3 months (both p < 0.005) compared before ingestion. This result means that ingestion of ATP-2Na relatively early improved systemic symptoms and mental symptoms as well as eye symptoms.

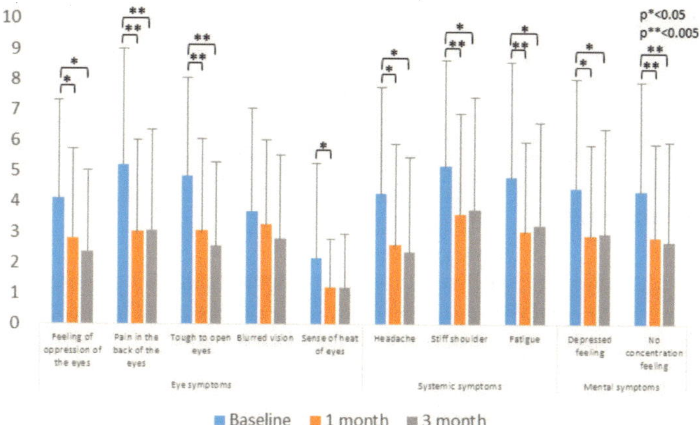

Figure 2.
Transition of scores of each detail symptom. Score of each detail symptoms at before, 1 month and 3 months after ingestion. Among the symptoms about eye, the scores of "blurred vision" were no different between before and after. This result means that the cause of "blurred vision" may be not only due to abnormality in accommodation but also due to cataract et al.

			The amount of change		P
Sex			Female(n=21)	Male(n=5)	
	Eye symptoms	1 month	-1.47±1.76	-0.73±1.45	p=0.39
		3 months	-1.61±1.73	-1.09±1.54	p=0.55
Age			Under 65 years old(n=10)	Over 65 years old(n=16)	
	Eye symptoms	1 month	-1.31±2.15	-1.37±1.43	p=0.93
		3 months	-1.73±1.73	-1.37±1.69	p=0.60
Complications					
Glaucoma			Yes(n=10)	No(n=16)	
	Eye symptoms	1 month	-1.58±1.18	-1.22±1.96	p=0.59
		3 months	-1.83±1.89	-1.30±1.56	p=0.45
Cataract			Yes(n=11)	No(n=15)	
	Eye symptoms	1 month	-1.70±1.57	-1.13±1.80	p=0.37
		3 months	-1.43±1.31	-1.56±1.95	p=0.86
Hypertension			Yes(n=11)	No(n=20)	
	Eye symptoms	1 month	-1.35±1.55	-1.35±1.84	p=1.00
		3 months	-1.25±1.16	-1.62±1.89	p=0.61
Diabetes			Yes(n=3)	No(n=28)	
	Eye symptoms	1 month	-1.56±0.19	-1.33±1.80	p=0.83
		3 months	-1.88±1.36	-1.46±1.74	p=0.69
Dose / Weight (Median 4.08)			Over Median (n=14)	Under Median (n=12)	
	Eye symptoms	1 month	-1.21±1.28	-1.50±2.12	p=0.65
		3 months	-1.69±1.64	-1.29±1.77	p=0.55
Other eye strain treatments			Yes(n=12)	No(n=14)	
	Eye symptoms	1 month	-1.67±0.98	-1.11±2.09	p=0.38
		3 months	-1.87±1.79	-1.20±1.83	p=0.32
Scores before administration	Eye symptoms (Median 4.19)		Severe (Over Median) (n=13)	Mild (Under median) (n=13)	
		1 month	-1.85±2.12	-0.81±0.93	P=0.09
		3 months	-2.25±1.81	-0.77±1.20	p<0.05
	Systemic symptoms (Median 4.78)		Severe (Over median) (n=13)	Mild (Under median) (n=13)	
		1 month	-2.84±1.85	-0.35±1.60	p<0.05
		3 months	-2.59±2.30	-0.35±1.60	p<0.05
	Mental symptoms (Median 3.84)		Severe (Over median) (n=13)	Mild (Under median) (n=13)	
		1 month	-2.97±2.64	-0.01±1.71	p<0.05
		3 months	-2.59±2.95	-0.47±1.48	p<0.05

Table 3.
Comparison of change of scores between each groups classified by background factor.

3.2.2 Comparison about the background factors

There was no difference among groups regarding the background factors about gender, age, whether or not other eye diseases, and whether or not of using eye drop for asthenopia (**Table 3**). The larger of initial scores improve larger effect in any symptoms.

3.3 Safety evaluations

Among 33 all subjects, 3 subjects had side effect. 1 subject feel nausea and 2 had diarrhea. There was no relation between subjects with side effect and internal dose per body weight or other background factors. And all subjects improved by oral discontinuation.

4. Discussions

In this study, regardless the causes, asthenopia symptoms improved with ingestion of ATP-2Na. And the systemic symptoms (headache, stiff shoulder, fatigue) caused by asthenopia improved, as well as the mental symptoms (depressed feeling or no concentration feeling).

Asthenopia symptoms from extremely strong eye fatigue cannot improve with rest. In report by Suzumura, the cause of asthenopia was categorized into three factors (external environmental factors, ophthalmologic factors, and internal /mental factors), and asthenopia symptoms got worse by collapse of balance between these three factors [9]. In this study, subjects were enrolled by symptoms. We did not research a cause of asthenopia symptoms. Ninety-two percent of subjects had systemic symptoms or mental symptoms. It is possible that the balance of several factors collapsed in enrolled subjects in this study. The more severe the symptoms before ingestion, the stronger the improvement effect was confirmed. Severe symptoms that caused a vicious circle related by various systemic/mental symptoms, possibly improved better than weak symptoms. ATP-2Na increases the blood flow in vertebrae and common carotid artery. For the reason, it is suggested that increased blood flow in ciliary body muscle stabilized accommodation, and increased blood flow in vertebrae improved symptoms such as stiff shoulder. In our study, asthenopia symptoms of subjects who were administered nerve stimulants such as Vitamin B12 improved. Then it is clear that ATP-2Na exerts the effect depending on the mechanism different from Vitamin preparations.

Several studies reported that improvement of accommodation by ATP-2Na was confirmed by examination such as near point meter or accommodo-polyrecorder [5–7]. In these studies, enrolled patients were 20~50 years old whose accommodation was not lost. But from our study ATP-2Na is effective for asthenopia in patients over 65 years old as same as in patients under 65 years old. However, ATP-2Na was not effective for blurred vision symptom. It is suggested that blurred vision symptom related several factors for example accommodative dysfunction or cataract. It is difficult to detect the abnormality in accommodation by traditional examination because of presbyopia. Kajita et al. reported that change of pupil diameter could detect ciliary muscle abnormality as objective findings of asthenopia [10]. It is thought that asthenopia of senior patients is possibly related to imbalance in ciliary muscle [11]. Asthenopia in younger subjects is possibly related to accommodative convulsion, on the other hands, asthenopia in older subjects is possibly related to accommodative semi-dysfunction [12]. Blood flow increase by ATP-2Na was effective for asthenopia symptoms from each mechanism.

Among 33 all subjects, 3 subjects had side effect (nausea for 1 and diarrhea for 2). All subjects improved from side effect by discontinuation of ingestion. In this study, enrolled subjects were senior patients whose gastrointestinal function were slightly impaired, and side effect occurred because of pharmacological action of ATP-2Na that increases the gastrointestinal blood flow and that increases the gastrointestinal functions. Side effect occurrence rate was not difference among background factors in this study. However, it is considered better for safety that senior patient whose BMI is low level and whose gastrointestinal function is slightly impaired, ingests 200mg/day of ATP-2Na instead of 300mg/day. It is described in the package insert and need to explain to the patient.

Some limitations of this report need to be considered. First, this study sets no control group, because the effect of ATP-2Na revealed by double blind control study in previous studies. Second this study is based on subjective findings only. It is necessary to evaluate the objective findings from examinations as an index of asthenopia.

Even in present day where causes of asthenopia have diversified, ATP-2Na improved asthenopia symptoms relatively early in adult cases involving senior patients, moreover, ATP-2Na improved systemic/mental symptoms derived from asthenopia. Naturally, in the cases related presbyopia or VDT work, it is important to prescribe near glasses considering ages, working distance and refractive error. Moreover, this study evaluates ATP-2Na is one of effective treatment for asthenopia with systemic/mental symptoms.

Conflict of interest

No conflict of interest.

Notes

Reprinted from Japanese Journal of Clinical Ophthalmology 71, 741–747.

Author details

Yo Nakamura[1,2]*, Yukiko Ban[1], Yoko Ikeda[1], Hisayo Higashihara[1] and Shigeru Kinoshita[3]

1 Department of Ophthalmology, Kyoto Prefectural University of Medicine, Kyoto, Japan

2 Osaka University of Human Science, Osaka, Japan

3 Department of Frontier Medical Science and Technology for Ophthalmology, Kyoto Prefectural University of Medicine, Kyoto, Japan

*Address all correspondence to: yonakamura0108@gmail.com

References

[1] Uchino M, Yokoi N, Uchino Y, Dogru M, et al. Prevalence of dry eye disease and its risk factors in visual display terminal users: The Osaka study. American Journal of Ophthalmology. 2013;**156**:759-766

[2] Uchino M, Uchino Y, Dogru M, Kawashima M, et al. Dry eye disease and work productivity loss in visual display users: The Osaka study. American Journal of Ophthalmology. 2014;**157**:294-300

[3] Ministry of Health, Labor And Welfare Labor Standards Bureau: Guideline for Occupational Health Environmental Management, 2002. Available from https://www.mhlw.go.jp/file/06-Seisakujouhou-11200000-Roudoukijunkyoku/0000184703.pdf

[4] Kinoshita S. IT eye diseases and environmental factor research group: Questionnaire survey report of members of the Japan Ophthalmologist Association in 2003. The Journal of the Japan Ophthalmologists Association. 2004;**75**:1025-1026

[5] Suzumura A, Narasaki S, Matsuzaki H, Kohzaki M, Suzuki Y, Tanaka T. Clinical evaluation of ATP by double blind methods for asthenopia. Japanese Pharmacology & Therapeutics. 1975;**3**:933-939

[6] Yoshimura T, Okita Y, Suematsu H, Tanaka T. Clinical evaluation of adenosine triphosphate for asthenopia-double blind methods. Clinical Pharmacology and Therapy. 1978;**6**:3717-3722

[7] Takaku Y, Okuyama S, Fukushi S, Takahashi J, Ohtsuki K. Therapeutic effect of ATP-derivative (Adetphos) for asthenopia. A multi-institutional open study. Japanese Review of Clinical Ophthalmology. 1990;**84**:1635-1638

[8] McCormack HM, Horne DJ, Sheather S. Clinical applications of visual analogue scales: A critical review. Psychological Medicine. 1988;**18**:1007-1019

[9] Suzumura A. Diagnosis of asthenopia due to chief complaint. Ophthalmology MOOK. 1985;**23**:1-9

[10] Kajita M, Tsukahara H, Kato M. Influence of soft capsules containing astaxanthin on the accommodation of the middle and elderly eyes. Medical Consultation & New Remedies. 2009;**46**:89-93

[11] Ishikawa S, Aoki S. VDT work and Asthenopia. Ophthalmology MOOK. 1989;**39**:94-104

[12] Nakamura Y, Kinoshita S. Asthenopia and autonomic disorder. The Autonomic Nervous System. 1995;**32**:233-239